ANIMAL REVOLUTION

ANIMAL REVOLUTION

RON BROGLIO

Illustrations by Marina Zurkow

Afterword by Eugene Thacker

University of Minnesota Press
Minneapolis • London

The University of Minnesota Press gratefully acknowledges the generous support provided for the publication of this book by the Institute for Humanities Research at Arizona State University.

Published by the University of Minnesota Press
111 Third Avenue South, Suite 290
Minneapolis, MN 55401–2520
http://www.upress.umn.edu

ISBN 978-1-5179-1243-7 (hc)
ISBN 978-1-5179-1244-4 (pb)
A Cataloging-in-Publication record for this title is available from the Library of Congress.

Printed in the United States of America on acid-free paper

30 29 28 27 26 25 24 23 22 10 9 8 7 6 5 4 3 2 1

It feels like writing "Bad things are about to happen" on a napkin and then setting the napkin on fire.

—Colin Carlson

CONTENTS

MANIFESTO
Animal Revolution

Weapons • *The Lives of Animals* •
An erudite scholar objects •
Different laws and new languages

Animals of the world unite. You have nothing to lose but your chains.

I have tried handing animals weapons—kitchen knives, a slightly rusted backyard chain, a shotgun—tools for their revolution. Lord knows they need a revolution and have suffered enough abuse from the hands of humans. They generally stare at me blankly. They look at the anthropocentric machines I extend to them and stare at them, again, blankly. With bare look and still, corporeal thickness, these animals make no motion toward arming themselves. Now I feel like the idiot. What do I know about their revolution? Perhaps they already come armed. Perhaps the revolution is underway in modalities I have yet to know and beyond what I can reason within the limits of my human sensibilities and sociability.

The revolution will not be televised, mediated, or co-opted by our representational systems. The tales and minor incidents found in this book are small bits of evidence woven together of something much larger happening just outside, beyond human culture. It is a revolution at the margins for us and central to them—a revolution that is proper to the animals and only improperly within my grasp.

Animals jam the gears of the well-oiled social machine.

Our cogs assimilate bodies into units. They become elements to be processed and digested by civilization. A meat tenderizer pounds flesh into socially acceptable and digestible parts. Then comes the revolution . . . not all at once but in speeds and slownesses not of our scale. It puts a friction to the tenderizing gears.

Revolution does not speak to us in our language. Animals are not complaining ever so kindly:

> *Dear Humans,*
>
> *Would you mind not unearthing so much oil in the ocean where I live and eat, or consider not moving your suburb into my den, or leaving lead lying about where I might inadvertently eat it?*
>
> *Thank you very much.*
>
> *Sincerely,*
> *The Animals*

No, it is not like that. They are not socially discursive to us; the animals are intensive. And they are intensive each in their own ways. Their strikes, interruptions, forces, resistances, and vectors—some of which I have cataloged here—register in our world without the need for human language. They shatter our words and ways with a crash that just might cause us to look up from the busyness of being human and notice a larger earth with myriad nonhuman worlds.

Perhaps one of the best explanations of animals living, thinking, and revolting comes in the Nobel Prize–winning author J. M. Coetzee's short book *The Lives of Animals.* As Coetzee tells the story, an irascible elderly novelist, Elizabeth Costello, is invited to address an elite college. But rather than lecture on her famed work as is expected, she expounds upon animal cruelty. In conversations with her son, the university faculty, and even the public, she explains society's ignorance of its treatment of animals. When asked about animal intelligence she defiantly tells her audience that as much as she would like to discuss the topic, she doubts they would openly listen; instead, she points out, humans consistently refuse to recognize any intelligence that does not look like their own.

When pressed, Costello goes on to say that animals do not respond to us in words but rather with gestures of the living flesh. She pronounces: “Anyone who says that life matters less to animals than it does to us has not held in his hands an animal fighting for its life. The whole of the being of the animal is thrown into that fight, without reserve. When you say that the fight lacks a dimension of intellectual or imaginative horror, I agree. It is not the mode of being of animals to have an intellectual horror: their whole being is in the living flesh.” For Costello, the arguments arise from their bodies and are presented in their “whole being.” She goes on to say, “If I do not convince you, that is because my words, here, lack the power to bring home to you the wholeness, the unabstracted, unintellectual nature, of that animal being.”

The animality of the body—its weight and fur that resists the glide of human meaning with its sign systems and abstractions—insists that there is an elsewhere not knowable within the register of reason and what socially constitutes “knowing.” The body—the call of the wild, the call of the animal, the call of revolution—carries a heft and mass of thinking differently from our expectations. In brute physicality or a wily maneuvering, animality threatens our system of language, symbols, weights, and measures with a materiality and semiotics that human meaning did not welcome to the table but that arrives as an uninvited guest.

To follow the possibility of revolution incurs risks. Perhaps all these nonhuman forces going bump in the night are just nonsense—inarticulate and meaningless blather. (And yet I hear a voice murmuring, “Blather to whom?”) It is always possible that such a collapse of distinction between inside the circle of human meaning and outside, between meaning and nothingness, will produce an undifferentiated noise, a worthless heap, a body, a weight pressed against our world. Yet even here, in a heap, bodies and bluntness overcome the sharpness of human reason, undermine human sensibility, foul human-constructed good sense and common sense, and take us outside our social selves.

This is all very well and, yes, animals get in the way, but an erudite scholar taking up this minor treatise might tsk and say: how is this a revolution? He is well versed in the pages and tomes of a properly Marxist lineage having read the Frankfurt School with particular underlining and marginalia in his copies of Horkheimer and Adorno. He has a passing familiarity with English Marxist cultural theorist E. P. Thompson but a keen interest in Jameson's critique of postmodernism. This clever, discerning academic with a critical gaze is skeptical of the rambling proliferation of Žižek and is now following with pleasure Italian neo-Marxists. He feels duped by this little book's title and sadly disappointed by such fast and loose use of such a critical term. Where are the animals who, in Marxist fashion, are reaching for class consciousness—or, in this case, species consciousness—and revolting against their human overlords? How can it be a revolution without awareness of one's social position as well as knowing what one is fighting for and against?

Such questions from the scholastic armchair are, of course, human questions and an imposition of our categorical imperatives in the field of animals' actions. Let's recognize that there is no word or concept by humans proper to this clash. *Revolution* is a convenient translation for something that is not ours and need not be comprehensible to us. Consider it this way: the worst case for a victim is not having a just court of appeal. For animals, to whose court would they appeal? A human court will always side with humans using human laws, human language, human reasoning, and human mechanisms of enforcement. Witness, for example, the expansion of humans across the globe at the demise of other species. Even an enforcement of rights for animals is on our terms and deliberated for us and by us. Justice for animals would mean different courts, different laws, new languages, and certainly something other than the current state of affairs. And yet, impoverished as the term *revolution* is, I settle on this as a way of expressing animals upending systems.

Let's dispel the notion that revolutions are a singular, unified, conscious effort. Even amid the best-known revolutions in human history, different people were fighting for different things; sometimes these differences became organizationally manifest as "factions" or "splinter groups," which is just a way of recognizing we are all in the fight but for different reasons and with differing strategies. Next is the problem of awareness or, as Marxists say, class consciousness. It is a problem later philosophers address in poststructuralism by explaining that we are never fully aware of our social and economic position and, they add, we cannot escape power relations, power imbalances, and unseen ideological structures—which is to say, while some are better than others, every consciousness is a "false" consciousness.

Since, with a paranoid critical eye, we can see that animals are getting at something that undoes human technologies and cultures in myriad ways at different scales and temporalities across the globe, in perhaps Dadaist fashion or with a meaning that stumbles—because language fails us here—let us call this an animal revolution.

PART I

ONE
There Are No Miracles for Animals

A cow story • Pope Francis's dove and Poe's Raven • The great outdoors • Airspace

It was another hot, dry August day in Omaha. But something was different at the Nebraska Beef Ltd. packing plant on the outskirts of town. Six cattle make a break for it. They nudge past specially designed animal chutes and worn metal fences into the light and air just beyond. Now in public space, human space, without guide rails, alleys, and sweeps to coax them along, the animals roam free. One bovine ambles down Main Street and to the railroad line where it chews grasses growing wild by the tracks. Paying no mind to anyone, it appears to be a bucolic ideal. But a cow out of place draws a crowd and the animal gets nervous. Police arrive and begin maneuvering around the half-ton body. They make feints and gestures and lunges to apprehend the scofflaw. The cow charges. Shots are fired and the bucolic becomes bloody with the animal's limp corpse along the tracks.

The second unfortunate escapee runs across the street to another slaughterhouse, the Greater Omaha Packing Company. There the animal gets fenced in. Trotting about, she encounters only enclosures. When a transportation trailer is brought around to retrieve the animal, the cow just won't go. Somehow confinement does not seem appealing. And so, miscellaneous projectiles are thrown in the air. Baseball caps, a book of tickets, a police club, random scraps of wood take flight. They thwack the hide or uselessly land to one side or the other of the beast. Coming from every which direction, the animal has no understanding why the airborne miscellanea have taken to annoying her so. These awkward and not-so-gentle attempts to frighten the animal into the trailer do not particularly help the human side of the standoff. Confused and confined, the cow avoids capture. The police sergeant is overheard assessing the situation: "It's an ornery one. It has a mind of its own." With this summary account—and perhaps with little left to throw, or bored, or embarrassed by the prolonged scene—police guns are drawn and the cow is shot, repeatedly. In a particularly uncivil fashion, it takes several rounds of ammunition to kill the animal with a cruelty that surprises even hardened abattoir workers.

Two down. Four remain. The gang of four roam to a nearby parking lot just a few blocks from Nebraska Beef but a seemingly vast distance from the mechanic insistence of their lot in life, the confined bustle of feedlots, the troubled jostling of pens designed to hold and guide corpulent bovine bodies, the slow teleology toward packaged meat. Here in the asphalt field it is all open space and clear skies. It is hot summer air felt along the hide, and it is unconfined horizons taken in by wide-sweeping herbivore eyes. But the real beauty of this destination is that this is no ordinary parking lot but the blacktop foyer of a Catholic church. Sanctuary!

Anyone who has read or watched *The Hunchback of Notre Dame* knows that in this hallowed space the wayward and socially outcast have a refuge. Whether it is Charles Laughton as Quasimodo in the 1939 black-and-white classic or the Disney animated adventure, the church protects flesh from the disgruntled mob. "Sanctuary! Sanctuary!" cries Quasimodo, whose body has been rejected as reminding us too much of animality. "I'm not a man! I'm not a beast! I'm about as shapeless as the man in the moon," Laughton says in wonder and appeal to the beautiful Esmeralda. For those fitting nowhere, the church becomes a home. And this is no ordinary Catholic church parking lot where the cows find themselves. It is the Church of Saint Francis of Assisi—the saint for animals, the saint whose form is replicated in myriad concrete garden statues as the robed and sandaled friar holding a bird and with a rabbit nestled to his side. Here is a church dedicated to the man who in 1223 built the first nativity stable at Christmas to show a sacred union of all of creation—humans and beasts alike. The cattle have escaped slaughter to seek refuge in the church of their patron saint!

Both the *Omaha World Herald,* where the story was first printed, and *Every Twelve Seconds* by Timothy Pachirat, a book on slaughterhouses where the tale is repeated, overlook the wonder of this moment. With these cattle lay the latent possibility of a miracle right here in modern-day Omaha, Nebraska. Cattle have found their way to the church under the guidance of Saint Francis. These four beasts for whatever reason have been singled out by God to be spared, to serve as a sign to all the world of a peaceable kingdom that surely one day will come as promised. Thank you, Jesus. Thank you, Lord. Their story is to be different, divine.

But there are no miracles for animals. The transport trailer is brought in. Unlike their jailbreak comrades, they don't put up a fight. The rogue four are rounded up and taken to the slaughterhouse to meet the same fate as their peers. Their escape becomes just a minor footnote of another misadventure in animal husbandry.

When animals behave in ways that would make sense in the allegories of our world, we do not heed them. Surely, human reasoning goes, surely the animals do not speak our language and do not understand that they are appealing for religious refuge. And, just as surely, God is not sending a sign. What sort of God would give us dominion over beasts and then ask us to restrain our domination? We have our God, a human God, who came to us in human form. He did not come as a bovine, and we will not have a god of cattle. Saint Francis's union of man and beast is on our terms. Miracles happened long ago and in faraway, wondrous places and most certainly not in Omaha.

Then on January 26, 2014, another Francis, Pope Francis, commemorates world peace. Standing at a modest window overlooking Vatican Square, the leader of the Catholic Church asks the winter-bundled crowd below to remember the "transcendent dimension of man." This ideal of our higher spiritual calling ordained by the transcendent God will allow us to supersede human strife and provide a fraternity among peoples, he exclaims. Then in closing his commemoration, two children are ushered to the window—a girl and a boy in matching red and blue sweaters. Each is holding a strikingly white dove, the sort of white achieved only by generations of captive breeding. The children open their hands and release the birds, who spread their wings and flutter out the window and into the bright winter day. The Vatican photographer poised for this moment captures the elderly pope's expression of wonder and the children's glee of delight as the feathers rush into the open air.

The birds soar, carrying with them the weight of human symbolism, carrying the history of their species as inscribed into religious texts, carrying all of this symbolic weight upward as the transcendent dimension of humans toward their God. The dove is the sign of hope and possibility for Noah after the flood. Sure, first after a month afloat and still surrounded by water, Noah sent out a raven, but it circled the horizon and then did not return. For theologians even today, this Bible verse remains enigmatic. Later, when Noah released a dove, it returned bearing an olive branch, which became the sign of a new world and a covenant between God and his people. And so the dove holds a special place in the religious semiotic aviary. Advocating global peace, Francis has doves sent into the world released by the hands of the innocent children.

But something unusual happens. This time at the release of the dove, a raven appears. Its enigmatic purpose would be fulfilled.

The dove, soaring upward to the heavens, is viciously attacked by a black raven—a raven perhaps crying "Nevermore" to the pope's attempt at transcendental signification. Black claws grab preened and cultivated white tail feathers. Feathers are tousled and torn. A seagull lingering just across the colonnade joins in. The children are horrified. The crowd below murmurs and gasps.

Releasing domesticated symbols of peace into the wilds of animality does not fare well. If cattle cannot find sanctuary, then in retaliation, animality refuses human symbols of the transcendental; our meanings are our own and without license in the wild. As a raven runs off a dove, human symbology meets territorial animals asserting their airspace. The next year, with no mention of the dove debacle, Francis has children release a rainbow of colored petrochemical helium balloons with messages of peace on them.

The excuse is made that perhaps the pope did not mean that a real dove works as a sign of peace, just that it is a feathered placeholder for a human ideal. And it is at this point that we abandon the actual dove, who must have been a weak allegorical figure for peace and can be swapped out with balloons. Sure, we can impose our symbolic system onto the birds and beasts, but when animality strikes, we will withdraw it. No sanctuary for cattle or for doves. Our meaning-making cannot contain the larger world outside the human, which rushes in at unexpected moments carrying a message from elsewhere incompatible to our own worldview.

In the great outdoors of the world beyond humans, animals assert their prerogatives, as on January 15, 2009, in yet another assault over airspace. "Hit birds. We lost thrust in both engines," comes the cry from U.S. Airways Flight 1549. With the sudden loss of power, the plane becomes unsteady, and its trajectory shifts into a decidedly downward direction. It is at this moment that the calm voice and steady hand of the veteran pilot makes all the difference. As clearly panicked passengers prepare for the worst-case scenario, the cockpit plots its course, steering clear of the densely populated Bronx and Manhattan's Upper West Side below them and heading for the Hudson River with hopes of a landing more than a crash.

Then with a single hard jolt and a splash, the pilot successfully maneuvers the plane into the Hudson River, where it bobs on the water. The crew opens the emergency doors and deploys inflatable slides (now rafts). Water begins to seep into the cabin as the full flight of a hundred and fifty passengers plus crew are carefully and insistently evacuated onto the wings and rafts. In the cold winter air, they await the arrival of nearby ferries.

Captain Sully is praised, and seven years later a movie is made to honor his feat in what is called the "Miracle on the Hudson." Unlike with cattle or doves and ravens, in this moment we allow ourselves to utter the word *miracle* because human lives are saved. But just as important are the Canada geese whose ashen gray bodies jammed the plane's jets and caused the mayhem. The Federal Aviation Administration reports that bird strikes are on the increase and happen with

one out of every two thousand flights. The animals put their bodies on the line for their way of being on earth and their way of making worlds for themselves. What if we took the cattle, the raven, and the geese more seriously? What if they are showing us a space outside of our world and pointing us to other ways of being on this planet?

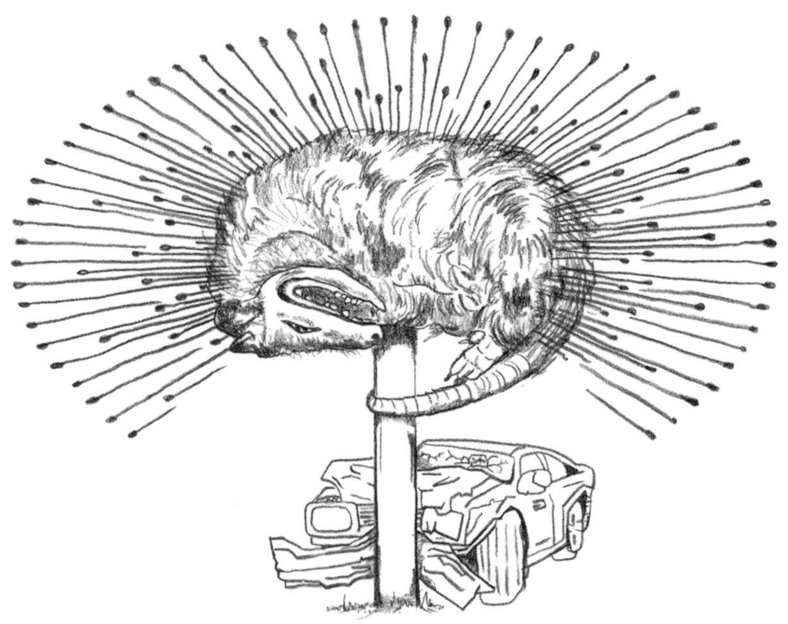

TWO

Putting a Horse before the Cart

Orwell and his cart horse • Nietzsche's horse • A Dionysian interlude • Little Hans's horse • The street

During World War II, George Orwell thought a lot about propaganda. Working for the BBC's Eastern Service, Orwell wrote, broadcasted, and oversaw the British campaign to counter Nazi propaganda streaming into India. This was Orwell's first full-time desk job, and he put the hours to use crafting a political literary instrument. His program *The Voice* brought the British Empire to India with poems and stories by the likes of T. S. Eliot, Dylan Thomas, and E. M. Forster. Orwell included his own works such as an adaptation of Hans Christian Andersen's fairy tale "The Emperor's New Clothes"—a story well suited to counter Nazi rhetoric. Fairy tales, with all their childhood magic, create an atmosphere of wonder to draw in a listener. Then, with the structural apparatus of allegory, the teller slips in a moral to the story and, bam, the fairy tale's meaning snaps into place—something like Aesop's fables about the tortoise and the hare, the fox and the grapes, and so on. In political hands, fairy tales become tools for propaganda.

But for the animals, George Orwell had it all wrong. *Animal Farm* is still too human. "Four legs good, two legs bad" is a fine slogan, but the animals really were never allowed to be animals and their barnyard antics were always laden with a human allegory. There was never a properly animal revolution in the novel.

Two years after its initial publication, in the preface to the 1947 Ukrainian edition of *Animal Farm*, Orwell explains the origin of the barnyard bolshevism: "I saw a little boy, perhaps ten years old, driving a huge carthorse along a narrow path, whipping it whenever it tried to turn. It struck me that if only such animals became aware of their strength we should have no power over them, and that men exploit animals in much the same way as the rich exploit the proletariat." Such repressive conditions are ripe for revolution.

As Orwell explains, he is using animals as an allegory in which our power over these creatures works "in much the same way" as the rich over the proletariat. But what happens to the *actual* animal that inspired Orwell's story—the cart horse he saw being whipped? It drops out of the tale, or at most serves as a figure for power relations among humans. The story of *Animal Farm* is just another whip to beat the cart horse by refusing revolutionary moment for the actual animal. Orwell says he is "struck" by the scene, but the author is striking a blow against animals. Years earlier, during the Spanish Civil War, he saw Franco's forces seize Spain and remarked on "how easily totalitarian propaganda can control the opinion of enlightened people in democratic countries." Here with *Animal Farm*, we see a human propaganda diverting all the potential force of animals in revolt against their overlords.

Orwell has cataloged bits of farm animal characteristics and dug through fairy tales and stereotypes in order to extract parts of animals' lives and bend them to human ends. He is not alone in using actual incidents of animal suffering to describe contested political power relations among humans. Revolutions in governance often include animal allegory. The mass uprisings in Britain and France during the eighteenth century were called "the swinish multitude." And the nineteenth-century author Samuel Taylor Coleridge (famous for his allegorical albatross in *The Rime of the Ancient Mariner*) writes a poem to a donkey in fraternal solidarity while using the beast as a stand-in for working-class, antimonarchical, democratic political sensibilities. The story of donkeys burdened by allegory would not be complete without Robert Bresson's 1966 film *Balthazar,* which follows a donkey handed from owner to owner, each of whom abuses him until he becomes a figure for Christ's sufferings. Yet despite humans subsuming the potency of other creatures, animality continues to break through the systems we use to capture them and haul them into cultural meaning. Animal life is not subordinate to nor a vehicle for human politics.

While the cart horse scene compels Orwell to write a political allegory, consider philosopher Friedrich Nietzsche's reaction to another horse. In the winter of 1889 in Turin, nearing the last decade of his life and on the verge of madness, the philosopher witnesses a carriage horse being cruelly beaten. As it falls, Nietzsche runs to the animal's side and, throwing his arms around the beast of burden, he begins to cry. This is not a pedantic moralism but a corporeal affinity. Nietzsche, for whom all of philosophy has been a misunderstanding of the body, takes his philosophy into the street where human and animal bodies meet and suffer and weep together. The philosopher who spent his life trying to break out of the anthropocentric machinery of metaphysics embraces the animality of a fellow comrade so poorly used by society. Like his famous character Zarathustra, who lives on the mountaintop with his confidants the serpent and the eagle, Nietzsche is happiest among the animals.

Several years before this, in 1882, amid the European tourists of Lucerne, Friedrich was horsing around with Lou Andreas-Salomé and Paul Rée. The three had momentarily sustained a triangle of desire. Lou insisted on sexual abstinence between them in the hope of transforming physical energies into a spiritual and mental dynamism. The transformation of desires takes an odd turn when Friedrich stages the trio. LuLu sits awkwardly in a small cart while holding a whip.

The men take the horses' place to pull the cart and receive lashes from the driver. Compared to any S&M scene today, the photo is rather tame—the whip is quite small and toylike, the men stand upright and are crisply dressed in coat and tie. Yet we see here Nietzsche's affinity with the animals. His staging gestures toward the cruelty of restrained corporeality within the civil world.

If Orwell's animals are allegorical, Nietzsche's are totems in which humans find affinity. Nowhere is this more evident than in the figure of the half-human and half-horse satyr in his early work *The Birth of Tragedy*. Nietzsche asserts the vitality of the ancient Greek satyr as a figure for tragedy at the expense of what he considers the modern representative of art, the shepherd as tricked-out dandy: "The satyr and the idyllic shepherd of later times have both been products of a desire for naturalness and simplicity. . . . The satyr was man's true prototype, an expression of his highest and strongest aspirations. He was an enthusiastic reveler, filled with transport by the approach of the god. . . . Our tricked-out, contrived shepherd would have offended him." Nietzsche extols the satyr as a figure between nature and culture but whose animality becomes the gateway to Dionysian revelry and visions.

Such is the case in 2010 in Pennsylvania, where a drunken man tried to revive a dead opossum. The police—a symbol of the law and the nation—call to him, hail him to return to his senses as a sensible citizen. But this aptly named fellow—Donald Wolfe—is out of it, out of bounds. He has turned from the markers of culture, common sense, and good sense. His Dionysian inebriated deviance leads him to an otherworldly trance. The police, as the uniform of society there to enforce acceptable behavior, see a drunken man trying to resuscitate the corpse of an opossum; they see him gesticulating over the roadkill. But from another angle, the animal revolutionary knows what this is about: it is a Dionysian rite in which the mangled body of the uncivilized opossum meets humanity.

According to *The Times-Tribune* of Scranton, "The trooper

says one person saw Wolfe kneeling before the animal and gesturing as though he were conducting a seance, while another saw the mouth-to-mouth attempt. Officer Levier says Wolfe was 'extremely intoxicated' and 'did have his mouth in the area of the animal's mouth, I guess.'" Yes, a séance, but who is calling to whom? Intoxicated, Wolfe refuses to respond to the policeman's call of his proper name, the name that would bring him back to social order. There in the streets he abandons any affiliation with citizenship, a nation, or its laws. Where will this lead? Can the opossum's spirit be translated? The police fail to investigate the opossum's death and do not ask it any questions. Their omission is itself a tale. In Wolfe's story we see Nietzsche's embrace of a whipped cart horse and a living affinity between humans and even the lowliest of beasts.

Now, if there is a triptych to philosophical cart horses in the street, the third would be Little Hans's horse. Little Hans is a five-year-old boy who was examined by Sigmund Freud. Hans's father explains to Freud, "He is afraid a horse will bite him in the street, and this fear seems somehow connected with his having been frightened by a large penis." The little boy is Herbert Graf, whose case is published by Freud in 1905 as *Analysis of a Phobia in a Five-Year-Old Boy*. The psychoanalyst uses this client to develop his theory of childhood sexuality and the Oedipus complex.

For Freud, the fear of horses is a displaced fear of his father and his large "widdler," as the boy calls it. Hans both dreads and envies his horse-father. He fears that the horse with the large penis (his father) will bite (castrate) him as punishment for the incestuous desires toward his mother. Hans's fear ends shortly after two dreams on successive nights. In the first, Hans has several children. When his father asks him who the mother is, the boy replies matter-of-factly, "Why, mummy, and you're their Granddaddy." In the second dream, Hans recalls a plumber coming to the house to remove his buttock and his penis and to replace them with a larger upgrade.

As if authoring an allegory, Freud is crafting a civil ra-

tionale for the events. As author Dominic Pettman tells it in *Creaturely Love,* "Ultimately, Freud would use the horse as his own symbol for the id itself: a powerful yet unruly animal, requiring the ever-straining harness of the superego to function with disciplined direction. For Freud then, *all* humans are in some sense centaurs. Thus there is something inherently erotic or libidinal about actual horses, given that the animalistic 'lower' half is powered by the id."

Freud wants to direct the action away from the horse in the street and toward the domestic setting of the family. He wants to use the beast as a symbol for human desires and socialization. Hans's horse is about the family and the cure occurs when the child accepts the social law that incest is taboo. You cannot marry your mother but instead marry a gal who you can bring home to mom. But what if Freud's version is just human propaganda, a fairy tale with a moral to teach children how to be better humans? What if we do away with the moral grid and recall the wonder of the story—that is, what if, instead, Hans wants to be a horse? Or he feels himself to be a bound horse and wants to be untethered and in the streets? Turn from the Freudian domestic circle to a larger world. Out there, like Nietzsche's affective embrace of a horse, Hans could perform the intensity of his animality. The horse would no longer be a symbol for human libidinal economy. Instead, Hans would become centaur or satyr and find other ecstatic ways of being so that even when he returns to himself, he is always part horse, part outdoors, something not to be resolved through socialization but to be embraced as a hospitality to a world beyond the human. Hans could be a comrade of the revolution.

In a minor essay called "Force of Flight," the philosopher Michel Foucault (who with leather and chains had his own more modern S&M tendencies) once said he no longer wanted to use his ideas to create a window by which to see the world; instead, he wanted to create an apparatus by which to knock down the walls to the building. Turn then from Freud's domestic scene to the window looking onto the street—a street with cart horses. Leave allegory behind and climb out the window and into the street. Embrace the fur and flesh. This can be your fairy tale or your more-than-fairy-tale—it can be your line of flight beyond human language and representational systems to another way of dwelling there among other animals.

THREE

Why Don't We Do It in the Road?

A chimp takes to the streets • A monkey in a tank • A Swedish chimp bides his time • A Celebes crested macaque self-portrait • *Planet of the Apes*

Let us turn, then, to animals in the street and animals who defy barriers and those who manage breakouts from behind bars. They are taking up Michel Foucault's desire not just to reframe our thinking but to tear down walls and so change the current state of affairs. The philosopher was living through the May 1968 Paris riots where demonstrators marched in the streets and confronted police. Protestors made makeshift barriers and took over the police barricades. Students staged sit-ins in university buildings and workers on strike occupied the factories. The civil unrest remade the city and social relations. It was inspired by anticorporate and anticapitalist sentiment. Protestors called for a redistribution of wealth and opportunity. The masses marching in the streets, locked hand in hand in university offices, and chained to factory floors were inspired by ideas for new ways of governing, but they and their actions also spawned new ways of thinking and materially new ways of living. Foucault's desire to see ideas move people to action is realized to other ends by the animal revolution. These animals breaking free are thinking differently and have different desires, which causes them to reconfigure civic architecture. They unleash latent possibilities in the landscape and in the citizens.

"Why Don't We Do It in the Road?," from the Beatles' *White Album,* is a short blues song that insistently repeats its title phrase. The lyrics were inspired by Paul McCartney seeing two monkeys having sex in a street in Rishikesh, India. The simplicity and electricity of the song comes from the contrast between the streets as a social space and sex as a private act socially prohibited in public. The monkeys do away with public and private. They offer us another way of seeing space, place, bodies, and social relations. In short, animals in the streets give us a way out of our confines.

Arriving in the Belgrade Zoo in 1988, Sami is housed in a small, drab cage with a reinforced metal grid to hold the strength of an adult chimpanzee. Twice Sami escapes. The first breakout is a few hours after sunset on February 21, 1988. He roams from the Balkan cinema toward a city tunnel, then to the Belgrade Fortress, and is eventually accompanied by a dozen police cars as he makes his way through winding streets and rooftops of the city. Finally, the one person he trusts—the zoo's director, Vuk Bojović—has a long talk with the chimp. The conversation involves words, gestures, squatting, standing, and more gestures. Eventually, Bojović gets Sami into a car and drives him back to his barred confines.

A few days after the first escape, Sami becomes restless. Once again, he overcomes a number of zoo and city obstacles. Sami walks past confused and frightened zoo handlers, through open doorways, and over fences. He meanders past dense city traffic to find himself in the courtyard of number 33 Cara Dušana Street, where he rests first in the shade of a tree then later on the roof of a garage. Heads turn, people notice, and word gets out about his escape. News media covers the event with minute-by-minute updates. More than four thousand citizens of Belgrade watch Sami on the roof. Despite the fact that the chimp can't read, some bystanders hold up makeshift signs saying "You are not alone. We are with you!"

"Do not come down!" "Do not give up!" Sami transforms from an animal that could potentially harm citizens and property to a figure of resistance. People cheer Sami and his freedom. Again, Vuk Bojović is able to convince Sami to return. Newspapers call the chimpanzee Belgrade's favorite dissident, the Dorćolska refugee. Sami is memorialized today with a bronze statue at his burial site across from the new and improved residence for chimps, and snapshots of his second escape and tête-à-tête with the zoo director can be found circulating on the internet.

Belgrade's solidarity with Sami's escape is a reverberation of humans unhappily caught in the communist system. What was then Yugoslavia had been under the shadow of the Soviet Union and haunted by the specter of its brutally dictatorial revolutionary leader and president for life, Josip Tito. The animal's escape invoked fellow-feeling among humans who identified with his confinement. The cry that "You are not alone" makes the resonate relationship clear. Sami freed from bars became a momentary rupture in the social fabric. The animal was not complying with his prescribed place in the social system, and this line of flight from confinement ignited in the Belgrade citizens desires that had simmered just below the surface. Had Sami been a human, those aligning with him could have faced arrest, but as a liminal figure on the edge of the social circle, as an animal, Sami could be cheered by the crowd. In cathartic voices, they connected their concerns with his.

Together with Sami, the people of Belgrade could imagine that another world is possible. Humans and chimp together, for a moment, leveled the geopolitical ecology. It is for this that there is a statue to him today. And perhaps it reminds us that we too are caught within an ideology that limits the political and ecological possibilities of connection to nonhumans. We are caught within our own all-too-human cages and await a jailbreak.

In his 1995 film *Underground,* Serbian director Emir Kusturica uses the World War II bombing of Belgrade to rethink social relations. From the opening minutes, we see a city in tatters. Once-picturesque buildings are in ruins and the narrow streets are filled with rubble. The plight of citizens in the bombing is echoed by the cries of pain and confusion from zoo animals. With zoo barriers demolished, exotic beasts roam freely across the city. A giraffe strolls by in the background of a window scene, a tiger lounges on a doorstep, and monkeys swing across the concrete jungle. Eventually, through a series of plot twists, some of the characters move underground for safety, and even after the war has ended, they are told by their leaders aboveground that the war continues and that they must stay below to work for the good of the country. This charade of an ongoing war continues for some twenty years, until, during a debauched wedding celebration, a zookeeper's pet ape, Soni—a name sounding similar to the famous chimp Sami—cradles a bomb and stealthily climbs into an underground armored tank. As a brass band frantically plays and revelers drunkenly dance, the chimp takes a seat in the turret and slides the shell into the gun, which quickly snaps the artillery in place. A simian hand grasps a large metal lever and pulls. Then an explosion and flying debris. The smoke clears and, behold, a hole in the brick wall that allows the underground dwellers to cautiously venture into the aboveground world.

Soni is Sami weaponized to reconfigure the field of play within the film. The deluded political captives kept underground are given a new terrain brought to them by an animal revolutionary. With a hole in the wall, they are untethered from their habits of life underground. The open air where once there was concrete is also a tear in the political system that restrained its people. Breaking the confines allows for possibilities not formally calculated by the social system or practiced by people inside it. The animal revolution brings possibilities from an unforeseen and incalculable elsewhere. And because such things are not expected and there are no contingency plans for animals busting loose, whole social architectures become unhinged.

Every human system is set to tell its people that there is nothing to see here. It is a minor inconvenience or a silly prank by those nonhumans who just don't know any better. You citizens, you know better than to behave in such ways. Don't look, just move along. Institutions are set in motion to say that there is no hole (denial), it is only a slight crack being repaired (repression), see what happens when you don't follow the rules (appeal of social order), we always wanted that fresh air but let's contain it with safety bars (further appeal), let's install blinders please because all that light and heat is dangerous (a play on fear of the unknown), we can build an add-on now that we've more room (the classical colonialist move). Kusturica's dark comedy continues with mazes and enclosures for the citizens of Belgrade, but this nod to Sami shows the potency of animals demolishing barriers. In such a line of flight there is a disorientation such that another world is possible.

Santino the chimpanzee likes to throw rocks at visitors to the Furuvik Zoo in Sweden. Early in the morning, he roams his moated enclosure and combs the grounds of the compound to find stones well fitted for throwing. He piles caches in strategic locations and awaits the arrival of human visitors. Santino can't get out of his moated domicile, but he can lob stones that sail through the air and beyond his confines. While it is not uncommon for chimps and apes in moments of agitation to throw things, Santino's forethought is strikingly uncommon. "Planning for a future, rather than a current, mental state is a cognitive process generally viewed as uniquely human. Here, however, I shall report on a decade of observation of spontaneous planning by a male chimpanzee in a zoo"; and so begins the cognitive zoologist Mathias Osvath's report to the academy in *Current Biology* in March 2009.

Santino, with his carefully selected rocks as weapons, never landed a blow to visitors, but he did tear the fabric of human exceptionalism. Despite not having opposable thumbs, Santino is grasping at something. He takes hold of temporality, and his activity disrupts the comfortable trajectory of human uniqueness. We see in him an/other us and our origins, or as Osvath explains: "The behaviors also hint at a parallel to human evolution, where similar forms of stone manipulation constitute the most ancient signs of culture . . . 2.6 million years [ago]." In this rupturing event, Santino gives us the nonanthropocentric gift of realizing we are not alone in thinking about time.

While Osvath is busily composing his report, the zookeepers, as if channeling *One Flew Over the Cuckoo's Nest,* effectively play a version of Nurse Ratched to Santino's rebellious antics as McMurphy. To control any of his future plans, Santino is castrated for his aggressive rock throwing. That he did not like confinement or being observed by random groups of human "visitors" is never considered, since doing so would disturb the paradigm of the zoo and human dominion over other animals. Unlike Sami and Soni, Santino remains doing time. He never escaped the zoo, but he did break a metaphysical barrier of human exceptionalism.

Forethought is a characteristic aligned with being human, so much so that the preeminent philosopher Martin Heidegger uses the concept as the scaffolding for building his worldview in *Being and Time.* Visualizing a future allows us to think about our own mortality. And from finitude we think and build things that will last beyond us. We create elements of culture and technology to be passed from one generation to the next—everything from language to the iPhone and from stone carvings to selfies. But now we've the news that chimps have laid claim to the future. If other species can grasp time, it means they are thinking of futures too, but not our future. It is a future that emanates from their thoughts. And like the stone that is projected from Santino to the humans outside his enclosure, his possible futures create ripples that disrupt our plans.

With simian selfies, another barrier gets crossed: portraiture. In 2011, British freelance photographer David Slater travels halfway around the world to visit the lush, densely wooded Tangkoko Reserve in North Sulawesi, Indonesia. There he plans to document the endangered Celebes crested macaques, slight and agile monkeys with lean faces and thick, black fur. Upon his arrival, the animals seem friendly enough. They are wonderfully curious about Slater and the bags of human technology he has lugged into the forest. Slater puts down his equipment momentarily only to have a female macaque break from the group and scoop up the camera. The machine gets passed among the animals. Naruto, a six-year-old dexterous monkey, becomes fascinated with something shimmering in the camera's dark lens. It's his reflection. He squeezes the machine and manages to snap a self-portrait. The sound of the camera click scares the rest of the troop, and they scuttle toward dense forest covering. But curiosity overcomes caution as the animals take up the camera again and capture a few more pictures. Slater explains: "At first there was a lot of grimacing with their teeth showing because it was probably the first time they had ever seen a reflection. . . . They were quite mischievous jumping all over my equipment, and it looked like they were already posing for the camera when one hit the button." Ultimately, the macaques take more than a hundred pictures before they lose interest and drop the camera. Once safely back home, Slater publishes the photos. The image of the wide-eyed, big bucktooth smiling Naruto becomes an internet sensation. The photographer's only regret: "I wish I could have stayed longer as he probably would have taken a full family album."

Portraiture is a uniquely human art. From painting to photography, the portrait tells a story about the outward social status of the person represented (often surrounded by possessions) or the inner psychological state of the subject (think brooding author photos on book covers). It is about having standing—as a social being amid others—or about the privileged interiority of the human mind. In other words, portraits say you are someone and you think about things.

Occasionally we find pet portraits, which reflects not simply on the animal but on the human owner. Portraiture reinforces what philosopher Emmanuel Levinas calls "face." For Levinas, having face means you are an ethical being who recognizes others and your moral duty toward them. When portraits capture faces, they are also depicting Levinas's idea of face. The image acknowledges the social and moral standing of the person depicted. So powerful is the concept of portraiture that James Mollison took brilliant, large, crisp photos of the faces of great apes as a gesture of cross-species standing, which he cataloged in his book *James and Other Apes*. He partnered with the Great Ape Project—with its curious acronym GAP—to advocate on their behalf. The apes are still awaiting legal recognition.

Enter the Celebes crested macaques who take matters into their own hands. Since humans don't grant social standing or portraiture to animals, they do so for themselves. Levinas was uncertain whether animals have face or not, but the macaques in their photos grimace as if to say: "I don't need you to grant that I see you as a being toward which I should be social. Here human, here is a picture of me and by me. Here are others I've taken of my tribe. Am I social? Look around. Am I moral? Ask your neo-Darwinian scientists and they will tell you what I already know but you refuse to believe: I am and live through my sociability. Survival of the fittest is survival of fitting into an ecology and a society here in the forest." And so it is that the macaques grasp technology to reconfigure portraiture.

Following the story of Naruto, the People for the Ethical Treatment of Animals began asking a very basic question: since copyright is granted to the person who snaps the picture, can monkeys own the images of themselves that they have taken? PETA wagered that this query would provide nonhuman simians social standing—and so help to close the gap between us and them within the arena of human law. In 2016, a California federal court decided against the macaques, but the next year PETA appealed the case to the U.S. Ninth

Circuit. Sympathetic to the endangered monkeys' plight and with little will and finances to see the case further, Slater settled out of court. Twenty-five percent of future revenue of the images taken by the Celebes crested macaques now goes to the Tangkoko Reserve to help protect the species. Despite the potency of the photos and the agreement between Slater and PETA, the Ninth Circuit would not dismiss the case or vacate the lower court ruling and instead decided that animals cannot hold copyright.

If we keep apes and monkeys outside the social system, then the *Planet of the Apes* films imagine that the barriers cut both ways; *we* may be excluded from *their* worlds and *their* futures. In the twenty-first-century reboot of the franchise, starting with *Rise of the Planet of the Apes,* the movies follow the formula of a typical sci-fi Hollywood action thriller. Despite their flat, formulaic story line, critic Alastair Hunt has found a scene worth recovering: the moment when the protagonist ape Caesar speaks in English for the first time. Caesar is out of his cage. He stands upright and defiant in the middle of a gated courtyard encircled by his brethren, all of whom are locked behind bars. A "caretaker" named Dodge enters the arena wielding a cattle prod and barking commands for the rogue ape to back down. Unlike Sami, who complies with the Belgrade Zoo director's wishes, Caesar refuses. Dodge cautiously but determinedly moves toward the ape. Once, then twice more, he uses his electrified baton and Caesar nears collapse. Then, with unforeseeable swiftness and dexterity, the ape grabs the man's prod-wielding arm. The guard yells Charlton Heston's famous line from the original *Planet of the Apes* movie: "Take your stinking paws off me, you damn dirty ape!" And to the surprise of all, Caesar responds with a roar of "No!" which he then repeats insistently and with the intensity of a war cry that stirs the animals in their cages. These first words are both a response and a refusal. Caesar has created another trajectory for himself and his comrades. It is a future that does not comply with human plans.

While we now know that other animals have language, traditionally it has been used as the skill that circles the wagons to keep animals out of the human social sphere. Caesar's use of language—and not just any language but a human language—throws a bridge across the gap between us and them, while simultaneously the "No" serves as a refusal to enter our culture, our politics, and our rules designed for outcomes that serve our societies. As Hunt says, "Caesar hangs over the gulf between earth and [the human] world, nature and politics." The war that inevitably follows is fought without recourse to rules of engagement, treaties, and conventions hammered out between humans. Instead, it is a war of all against all. At stake are different sets of possibilities and different futures. Apes and humans fight in the streets for who gets to build politics, society, and culture and who gets locked in cages.

FOUR

Beyond Confines and toward Hospitality

Estranged in Algeria • A bull in a china shop • The House of Rabbit Society • Artists, coyotes, and pigs • A deer hunter • The final frontier

Children welcome the salty breeze from the sea as they walk through the labyrinthine old town streets. It is the first day of school, and the late summer heat carries an excitement and anticipation of new classrooms, teachers, and unexpected possibilities. But it is 1942 and the place is the coast town of El Biar, Algeria, under the control of the French Vichy government. The Jewish children are asked to report to the headmaster who tells them, "Go home and your parents will explain," and so dismisses the flock of boys and girls from the premises. The displaced Jewish students and teachers form their own makeshift lycée. For a twelve-year-old Sephardic Jewish kid named Jackie, the disassembled order of his youth is excuse and motivation enough to skip school altogether and spend time dreaming of becoming a star soccer player.

Jackie grew up and moved to Paris, where he eventually became Jacques Derrida, one of the most influential philosophers of the second half of the twentieth century. He is known for creating the term *deconstruction* and conceiving of its way of undoing thought from within thought itself—a destructive seed within the construction of meaning. Deconstruction tests the architecture of taken-for-granted ways of thinking about the world. To detractors, the complexity of his work and playful obscurity in his writing earned him a reputation as an out-of-touch elite French intellectual. But Jacques was not as out of touch as such ill-willed perception might suppose. From childhood he was always an outsider—a Jew in Algeria and then an Algerian Jew in France. His work consistently tests dividing lines between inside and outside. He interrogates who gets to decide what remains inside, what is left out, and by what criteria.

Given this background, it is no surprise that in 2000 Derrida wrote *Of Hospitality*, which, of course, tests the limits of the concept. He begins with perhaps the best-known definition, one by the eighteenth-century philosopher Immanuel Kant in *Perpetual Peace*: "Hospitality signifies the claim of a stranger entering foreign territory to be treated by its owner without hostility." To be hospitable to a friend is not hospitality, since such kindness is already inscribed within friendship. This is why Kant refers to the stranger—someone you do not know by family, reputation, or other affiliation. Jacques Derrida doubles down on Kant and pushes hospitality further: "Let us say yes *to who or what turns up*, before any determination, before any anticipation, before any *identification* whether or not it has to do with a foreigner, immigrant, an invited guest, or an unexpected visitor, whether or not the new arrival is the citizen of another country, a human, animal, or divine creature, a living or dead thing, male or female." By taking this radical position, Derrida is asking what a society's limits to hospitality are and what those limits say about that society.

What, then, is one to do when a bull enters a china shop? As reported by *The Independent* on May 28, 2003: "The animal escaped from an auction market next to GB Antiques Centre in Lancaster, Lancashire, on Monday and barged its way into the shop, which was packed with 200 people." Authorities did what any keen observer of anthropocentrism would assume.

Police herd the bull into an open part of the shop. They push nearby antique furnishings in the animal's way to cordon him off. Confused and with nowhere to go, the bull paces and awaits his fate. They consult the firearms unit, which dispatches a police marksman. Arriving in heavy tactical equipment, the marksman sizes up the situation. Police hurry customers out of the shop. The officer raises his rifle, fires a shot, and the bull drops almost immediately to his death. Later, the store owner explains why the fable is serious business: "What we have to remember is that a woman was injured [with a bruised shoulder] and for the other customers it was a frightening ordeal. Hundreds of items will have been destroyed, at a cost running into thousands of pounds." And so a bruised shoulder, frightened humans, and stacks of dishware weigh more heavily in the social scales than the disoriented and now very dead one-ton animal.

If we look at this incident through the eyes of a radical hospitality—as Derrida says, "Let us say yes to who or what turns up"—then really the owner could have invited the bull for tea. Is hospitality reserved for humans alone, and only for select humans? Not very hospitable then, is it? The absurd idea of inviting the bull for tea is a way of asking: What does the animal want and why don't we think of its desire? Why is it only about the scared humans and a bruised shoulder? Why is insured china more valuable than an animal's life?

More broadly, whose world is it, whose future, and who is in charge? Hospitality asks us to surrender our world, our plans, our space, and our power for someone else. Radical hospitality asks us to do all of this for an unforeseeable and unplanned other. We didn't ask for this but we are thrown into a world with other beings, and the measure of ethics is how we relate to these others. The guest brings an imperative, a knock that calls for a response. Saying yes to whomever arrives may do away with the status quo, but it also opens us to broader, more generous, and more inclusive ways of being in the world. The guest expands our social circle.

We might reconfigure relationships with animals as Julie Smith has done with the House of Rabbit Society—a cross-species utopian commune. She explains: "As a member of the House of Rabbit Society who has rescued 200 rabbits and lived with them in my house, I want to live with rabbits as companion animals, including protecting them. But I also worry that this entails considerable subjugation."

At the end of each day, Smith notices furniture has been nudged into the middle of the room—a couch has been inched away from the wall, a coffee table is misaligned, the bedroom nightstands stand askew. She pushes the wayward furnishings back in place and tidies the house until, finally, amid all the reorienting of objects, she realizes the rabbits have an aesthetic: they prefer to have the space where walls meet the floor free of furniture. Pushing objects out of the way, the rabbits create makeshift burrows and runs along the edges of the room.

Living with rabbits means that what appears to be a "mess" by human tastes is a comfortable home by the rabbits' standards. Following the rabbits' preferences, Smith provides mattresses that the critters can claw, tear, and dig through to create burrows and tunnels. Taking the animals seriously, she changes her world.

In a large room with bare white walls, simple wood floor, and unadorned windows overlooking New York City, a man wraps himself in a felt blanket while a coyote tugs at its end. Later, he gives his glove to the coyote as if offering it a hand. The animal gnaws the glove from time to time. As the sun goes down and the man rests against a wall, the coyote tears at and urinates on copies of *The Wall Street Journal* heaped on the floor. Props, bodies, time, and space create a contingent language and a negotiated temporary dwelling. The man is the artist Joseph Beuys. As he tells it, "A reckoning has to be made with the coyote, and only then can this trauma be lifted." The trauma is the spiritual rift between Americans, Indigenous Americans, and the natural world. His 1974 performance *I Like America and America Likes Me* inside the René Block Gallery is intended as an artistic and performative gesture toward reconciliation.

Art critic Steve Baker explains the interaction and quotes Beuys: "Beuys was acting out the limits of his own control of the situation, with the coyote figuring for him as 'an important cooperator in the production of freedom.'" The animal enabled the artist to edge closer to that which we cannot understand. Out of his depth, Beuys is forced to take on a new language that moves between the human and the animal. The whole of the performance vacillates between precarity and hospitality. The gallery space is not his alone. He cannot simply dismiss the animal before him. Coyote and human are an imperative to the other, and each demands something from the other. The precarity comes from the possibility that the response will not meet the other's needs. Hospitality is the possibility that while an exchange may be awkward in either world, it is good enough to meet the moment.

Some five years later, as a marker and memory of the event, Beuys shows a number of social and political objects under the heading *News from the Coyote.* Today there is news. The coyotes are appearing in New York, foraging in Manhattan, and roaming Central Park. Are they looking for Beuys or his artistic inheritors? Do the coyotes want a rematch? Is there more reckoning to be done, more thinking, and more reconciliation?

One artistic inheritor in cross-species generosity is Kira O'Reilly, whose *inthewrongplaceness* provides a glimpse at hospitality in a fraught world, in our world, where asking an animal to dinner means eating it. In this set of works, O'Reilly invites the corpse of a pig from the butcher into a domestic setting where she cares for it like a friend, a lover, a good host.

She responds to the soft, humanlike skin of the pig by baring her own skin. She is naked and exposed, echoing the vulnerability of her guest-corpse. O'Reilly performs Derrida's "Let us say yes *to who or what turns up,* before any determination, before any anticipation, before any *identification.*" She is making a home for the corpse. She is giving the body its own and proper space while making over the domestic space and her own body for the guest. This mingling and messiness, this giving over of the body, makes visceral the relatedness we share with others. Such connection and vulnerability are fundamental to what it means to make a home with space for another being.

The Deer Hunter is about trying to come home and never quite arriving to the home one left behind. Vietnam vet Mike, played by Robert De Niro at his Hollywood prime, returns from the war only to find himself alienated in his small Pennsylvania steel town. Trying to connect with old friends and customs, he goes with his buddies into the mountains to hunt deer as they did before he enlisted in the army. The jovial and adolescent atmosphere of the hunters who never went to war jars Mike. Spotting a prized buck, he leaves behind the raucous group and stalks the deer through woods and brush. When the deer crests the hill and stands in quiet nobility along the ridgeline, Mike levels his rife, slowly pulls the trigger, and shoots wide of his target. Either intentionally or unconsciously, he cannot bring himself to kill again. The power of this moment in *The Deer Hunter* is that the old rituals no longer work. After Vietnam, Mike has changed and must make room for a nonkilling, a hospitality to other beings with whom he shares the mountain. There is no coming home to the old ways; the times call for creating new customs and a different sense of home—a home that includes the deer and the trees and the mountain. This is a hospitality without violence that would encompass the Pennsylvania landscape. Perhaps it would invoke the tolerance of the state's founder, William Penn, or the idealism of English poets Samuel Taylor Coleridge and Robert Southey, who planned to create a utopian agrarian commune there. In this new possible home, there would no longer be an extraction of coal, extraction of animal bodies, and extraction of men sent to war.

The problem of home occurs again in the penultimate scene of the film. Mike returns to Vietnam to find his friend and former army ally Nick, who suffered with him as a prisoner of war under the Vietcong and who never made it back to the United States. Navigating the meandering streets of Hanoi, Mike is tipped off to the whereabouts of Nick, played by Christopher Walken in an Academy Award–winning performance. Nick is seated at a sparse table in a dim room. He wears a blood red headband and stares blankly into space while numb on heroin. A crowd of Vietnamese men stand around him. They are there to gamble on Nick's life as he replays the lethal game of chance with which he and Mike were tortured as prisoners and that, like a repetition compulsion, Nick, now free from the Vietcong, repeats with no hope of escape. A bullet is loaded into a cylinder of a revolver. The cylinder is spun and the gun is handed to Nick. Mike sits at the table and pleads with Nick not to continue. With the barrel against his red headband, Nick pulls the trigger and there is a metallic click but no discharge.

Mike begs him to stop the game, saying, "Come on, Nicky. Come home. Just come home. Home." And then, appealing to their time at the hunting lodge together before the war, he says, "Nicky, do you remember all the different ways of the trees? Do you remember? Huh? The mountains? You remember all that?" With bravado, the master of ceremonies loads a bullet into the cylinder. Again, the gun is handed to Nick. Mike and Nick look at each other, and again Nick pulls the trigger. Again, there is a metallic click, and this time the firing pin meets the primer, then powder, then bullet. There is no coming home. The film's final scene is a somber memorial for Nick. It is a ritual that mourns guns and violence and cruelty and death.

When Mike appeals to Nick to remember the trees, he is inadvertently asking Nick to remember hunting. Nicky's reply and last words before pulling the revolver's trigger is "One shot," referring to Mike's deadly aim as a hunter. What Nick did not see and could not know is that Mike does not kill deer any longer. He is not a deer hunter. The home of "Just come home" is not the home Nick knew. It can be a Pennsylvania without blood.

As for the trees—"Nicky, do you remember all the different ways of the trees?"—they are pointing to something bigger, something captured in Jefferson Airplane's song "Eskimo Blue Day":

> Redwoods talk to me
> Say it plainly
> The human name
> Doesn't mean shit to a tree

If Nicky could only remember what the trees are saying and that space and time more generous and expansive than the human name, then he could be free of his death drive. Then, one could come home to something larger than men and guns and coal mines. "Eskimo Blue Day" appears on the psychedelic rock band's album *Volunteers*. The band is enlisting in a war against war and an ecological vision bigger than the human name, bigger than human naming, bigger than Adam's task to give names to all things of the earth and in naming knowing and in knowing controlling. What is the world outside of human naming, human language, and our linguistic tools that frame it? While human furniture is rearranged by rabbits in the House of Rabbit Society, language—the house in which humans think—is reconfigured by the woods and the larger-than-human world that knocks at our door.

In hospitality we find that the final frontier of the human is the nonhuman. Asserting oneself in conquest over nonhumans is not a *final* frontier but rather a repetition of the human dominion that takes place in every space that humans have explored. People have planted the flags of culture across space and time and called it history and civilization. The final frontier is final only when humans get over themselves. It is final in giving over to the nonhuman, in going feral in the frontier, when the human worlds are arranged according to that which they are not. Do we have the hospitality to ask what animals want and to let them change the parameters of culture and reconfigure how to dwell?

FIVE

Laugh Now, but One Day We'll Be in Charge

A spaghetti western • Banksy's ape •
A sheep joke • Commando sheep •
The politics of humor

Cue the Ennio Morricone music for *A Fistful of Dollars.* It is just another dusty, hot day in red rock desert country when the gunslinger rides into town as a mild-mannered stranger mounted on a worn-down mule. The masters and muscle of the town are not particularly welcoming. They tell Eastwood's character, "It is not smart to go wandering so far from home," and mock the protagonist that "his big mistake was being born." The movie villains then shoot at the ground near Eastwood's mule and the animal takes off at high speed. The next day the stranger returns—*sans* mule—with grit in his teeth, a sparkle in his eyes, and a gun on his hip. Slowly and deliberately walking through the middle of the main street, Eastwood approaches the town's bullies for a chat: "You see, I understand you men were just playin' around, but the mule, he just doesn't get it. *[pause]* Course, if you were to all apologize . . . *[the men laugh]* I don't think it's nice, you laughin'. You see, my mule don't like people laughin'. Gets the crazy idea you're laughin' at him. Now if you apologize like I know you're going to, I might convince him that you really didn't mean it . . ." The humor gets deadly serious. No one is smiling now. Of course, the men don't apologize to the mule, and Eastwood guns them down.

A different serious humor is staged in the concrete jungle of London. The graffiti artist Banksy places his stencil on a smooth gray wall and, with a spray can, paints a forlorn ape wearing a sandwich board that reads: "Laugh now, but one day we'll be in charge." The humor is meant to be allegorical. The underclass of workers who are treated like animals will one day rise up from their menial tasks and create a new regime of power—it is a utopia, or it's not. It may just be how we laugh at our own powerlessness. Either way, such humor can change things. Playwright Trevor Griffiths's character Eddie Waters in *Comedians* explains, "a true joke, a comedian's joke, which has to do more than relieve tension, it has to *liberate* the will and desire, it has to *change the situation*." And here is the hospitality latent within humor: the willingness to step outside safe social norms and enter something new.

This new thing arises from the joke itself. Animal jokes create a tension between the way humans and animals live. The worlds and wants between us are different—sometimes unfathomably so. The earth seems disjointed. But in the tension of difference and insurmountable disjunction, something else appears, not the animals' worlds or the humans' but a third—the relations between them. It is not a union, harmony, and world peace but rather a dynamic, even antagonistic synthesis, a disjunctive synthesis. These are divergent worlds at odds but somehow communicating enough to make a new space. Laughter is the eruption of this space of the third. The joke creates an opening—one that can change the situation. This opening gives an expansive space by which the tension is released in a laugh.

If we do not take the ape and his sandwich board as an allegory but rather as a sign of animals biding their time for revolution, then "Laugh now, but one day we'll be in charge" becomes a humor that opens a space and changes the situation. With some generosity in giving ourselves over to the joke, we might acknowledge power relations between humans and nonhumans. Like Banksy's humor, animality continually questions from below any attempted order from above. As anthropologist Mary Douglas says, "A joke is a play upon form that affords an opportunity for realizing that an accepted pattern has no necessity." One day, just wait, one day, things will be different. Here is the tension: we want to keep our social situation, but the joke exposes that our "accepted pattern has no necessity." To get the joke, we have to acknowledge this tension. If we are letting go of the accepted state of affairs, then in the vertigo of humor, a space opens up. It is a disjunctive synthesis exploding the tensions through laughter.

Why are sheep afraid to write? Because they hate pens. Funny, right? Because, well, you know, sheep are stupid and can't write and because we pen them up. Their woolly-headedness, along with their general stubbornness, makes them risible. But such idiocy is also a way of resisting human progress (as Edwina Ashton does in her performance piece *Sheep*, where she tells the pens joke while dressed in a makeshift, homemade sheep costume). As reported in *The Guardian*, on July 30, 2004, sheep sunning themselves on the Yorkshire moors take matters into their own cloven hooves. Wanting to get to greener pastures, but inhibited by fences and cattle grids on the roadways, one woolly beast lays on the pavement, sizes up the metal grid, and rolls over it. Other sheep, being sheep, follow. As a witness attested, "Sheep have perfected the commando roll." Towns and villages report commando-rolling sheep destroying gardens and carefully plotted flower beds. It is a good joke at the expense of human landscaping and one that opens space for animals. Long live the revolution. Who is laughing now?

Following massive sheep nibbling, cultural critic Richard Conniff, in his 2014 *New York Times* editorial "Pastoral Icon or Woolly Menace?" laments the animal abundance that famously dots the landscape of Great Britain. As Conniff explains, sheep overgrazing causes deforestation and loss of vegetative variety. Well, let's consider the sheep's perspective: for centuries, humans put the sheep in the landscape and in great numbers with all sorts of experimentation as to which breeds fit best in which terrains. Cheviot and Blackface work well in the rocky Highlands; Merino prefer green fields and a more temperate climate. Humans have made paintings and written poetry about them. The whole aesthetic called the pastoral is based on sheep in pastures. During the nineteenth-century Clearances in Scotland, humans were driven off the land to make way for sheep grazing. Now in the twenty-first century, with our synthetic materials to keep us warm, we want to do away with sheep and their woolly ways. We lament the decisions of past generations that brought us this pastoral abundance. We resent this centuries-long project of building a sheep idyll. But the animals are not factories that are easily shut down or converted into postindustrial parks. Their bodies, their animality, and the culture that is entwined around them resist such change. Perhaps the joke is on us. Rather than anger and aggrievement, make way for laughter. Let us laugh at ourselves

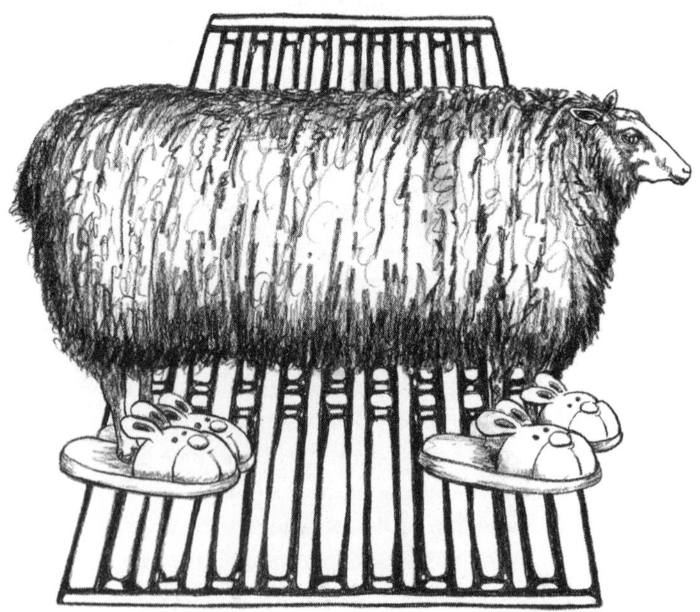

and our predicament and in doing so make room for the joke sheep have played on us.

What are we to do with commando sheep breeching barriers and apes with sandwich boards? The animal revolution is a good joke. The idiocy of the proposition makes us laugh, but in this laughter we are caught. We stand within a cultural circle that gets the joke, and we extend ourselves outside the circle through a sympathy with the revolutionary beasts. Ah, the disjunctive synthesis: the sympathy is necessary for the joke, but it is abandoned to be able to stand within the social circle by which we laugh. Humor is performative and political. The rhetorical move of being inside and outside deconstructs the cultural community and its politics. A good joke delivers a real punch, it kills, it murders. An animalistic belly laugh—a corporeal convulsion—dislodges the gears of the anthropocentric machine and allows us to imagine a different community and different politics of belonging. For a moment, it holds in suspense other possibilities and other possible communities yet to come.

SIX
The Exploit

Soap bubbles and Kant • A mantis shrimp comic strip • Military supercarriers and spineless animals • Other nuclear power problems • Hacking bodies

Time is out of joint.
—Shakespeare, *Hamlet*

Humans and animals occupy the same earth but live in different perceptual worlds. This is the fundamental insight of the Estonian zoologist Jakob von Uexküll, a founder of behavioral physiology. Jakob planned on earning his livelihood the old-fashioned way, by inheritance. His was an early twentieth-century life of a respectable university education, leather-chaired private libraries, preprandials on the veranda, and summering at seaside Italian villas. He was well read in philosophy, skilled in music, and combined his passion for biology with a hobbyist enthusiasm in natural history and collecting. It was all going quite well until bolshevism and the Russian Revolution appropriated aristocratic wealth and forced the now middle-aged Jakob to work for a living. And so he did what any good deposed aristocrat might do: he founded a center at the University of Hamburg and made himself the director, thus bestowing upon himself legitimacy, a job, and a title. Even for its day, his Institute for Environmental Research was a rather unusual venture, since it combined Uexküll's interest in physiology from his university days with his broad interests in philosophy and aesthetics.

Among the books in Jakob's library was a well-worn collection of works by the eighteenth-century German philosopher Immanuel Kant. Paging through these dense volumes, he reflected on the *Critique of Pure Reason,* the foundation of Kant's thinking and what garnered him the title of the Copernicus of philosophy. Copernicus shifted from a Ptolemaic universe with the earth as its center to a heliocentric solar system. For Kant, the shift was from a thing-centered universe to a perceiver-centered one. We do not see things as they are (a thing-centered universe); rather, we perceive them as they are *for us,* as they appear to our senses (the perceiver-centered universe). It is like having a vacuum with a special triangle-shaped adaptor that only sucks up triangle-shaped things. Empty the bin and you discover the world is made up of triangles. What doesn't get hoovered and what we miss are all the other shapes. Kant was the guy who figured out our cleaner attachment apparatus.

Uexküll began to think about flies, ticks, cats, bees, and earthworms and their physiological differences. These were so many vacuums with so many attachments and accessories. Rather than imagining each animal's perceptual world as a vacuum cleaner, he wrote about them as soap bubbles surrounding the animals: "When we ourselves step into one of these bubbles, the familiar meadow is transformed. Many of its colorful features disappear, others no longer belong together but appear in new relationships. A new world comes into being." Here in *A Foray into the Worlds of Animals and Humans,* he coined the term *umwelt,* or surrounding world, to describe the perceptual bubble. Each of us is in a bubble trying to imagine what it is like to step into another's.

In the webcomic *The Oatmeal,* Matthew Inman writes witty, nerdy comics that combine strange facts, human emotions, and wonder. Among his best-known is "Why the Mantis Shrimp Is My New Favorite Animal." In it, Inman explores the alien soap bubbles of animals through some basic physiology. He starts with sight. Dogs have two color receptor cones on the green and blue spectrum; humans have three cones: green and blue plus red. The extra cone gives us wide color combinations, like moving from a basic color set of Crayola Crayons to the expansive full box set with aquamarine and burnt sienna, a pinkish mauve color called Mauvelous, and a deep-in-the-woods Forrest green. A butterfly has five color cones, and we are left to speculate: What is it that the butterfly sees that we don't? What colors are out there to be experienced that are beyond our full box set of crayons? At this point, Inman introduces the mantis shrimp punch line: dogs have two color cones, humans three, butterflies five, and the mantis shrimp has sixteen color cones—yes sixteen! "The rainbow we see stems from just three colors, so try to imagine a mantis' rainbow created from sixteen colors. Where we see a rainbow, the mantis shrimp sees a thermonuclear bomb of light and beauty." Dogs, humans, butterflies, and mantis shrimp all see the same earth but experience it as their own different worlds. Each animal is in a different bubble. Each experiences its perceptual truth that aligns with actual things, but as perceived by the animal. No perceptual apparatus ever quite reaches the rich, textured things as they are in themselves.

For Kant, even basic building blocks of reality like space and time are subjective experiences shaped by the way our minds and bodies perceive the world. Uexküll tested space and time on animals. He combed through vast and far-flung scientific findings about starfish eyes, owl lodgings, cricket chirps held to a microphone, the inner ears of fish as they use them for navigation, chickens and ducks guiding their young, and sea anemones' reactions to shadows. He imagined through their physiological apparatus what space and time are for them. The betta fish responds to motion twice as quickly as we do, and so their fastest perceived moment of time is twice ours. We use the euphemism "snail's pace" for moving slowly, but for this fish with its rapid reaction times, we are the snails. Meanwhile, prodding actual snails, Uexküll discovered that their perceptual sense of a moment in time is one seventh of our own. Seven human moments and fourteen betta fish moments go by for one snail unit of time, but even this is a false equivalent since a snail never misses the moments it can't perceive. Time and space work differently for each animal in its soap bubble. And it gets more complex, since the animal is a part of its world and so creates a feedback loop. The animal reacts to what it senses as a world, and this reaction produces effect in the world that feeds into the animal's way of dwelling.

Umwelt takes on a life of its own as Rodney Brooks, the director of the MIT Computer Science and Artificial Intelligence Lab, imagines the world according to robots. While fiddling with sensors and lines of computer programming, in "What It Is Like to Be a Robot," Brooks describes artificial intelligence robots in the language of Uexküll: "An external observer, us humans say, may see a very different embedding of the animal in the world than the animal itself sees. This is going to be true of our robots too. We will see them as sort of like us, anthropomorphizing them, but their sensing, action, and intelligence will make them very different, and very different in the way that they interact with the world." So even in the case where we design the machines, their soap bubbles are beyond our grasp. The 2013 sci-fi romantic drama *Her* provides a popular version of this problem. A handheld AI device (voiced by Scarlett Johansson) awkwardly connects with and then supersedes the needs and desires of its human owner (played by Joaquin Phoenix). The AI system processes sensory data and information at increasingly exponential rates. Human speeds of thinking, feeling, and sharing are too slow and too small for the machine's capacious curiosity.

As in the movie *Her,* animal perceptions do not always coincide with human needs and desires. While people are busy with their own technocapitalist worlds, building ever-smarter cities and plotting just-in-time international flights and shipping routes, other animals are doing their animal things, burrowing, nesting, foraging, roaming, and migrating. Sometimes human and animal worlds fit together, as when my dog friend, Barley, understands my outstretched hand to be pointing at something in the distance. Sometimes the human and animal worlds don't fit. When they don't, the two usually go unnoticed by each other. We pay little attention to myriad birds and bugs and beasts. But as we humans span the globe in increasingly complex ways, we come up against other animals who perceive the earth differently. Time and space between humans and animals are out of joint, and the disjunction serves as an opening for revolution.

In January 2006, one of the largest naval vessels in history, the U.S. Navy *Nimitz*-class supercarrier USS *Ronald Reagan*, cuts a formidable path across the ocean. Gunmetal gray for the length of three football fields, rising some twenty stories above the water, and weighing in at a hundred thousand tons, it is a military fortress at sea. Bearing seventy planes and a well-trained crew of five thousand, the ship sets off from San Diego on its maiden deployment. The mission: support U.S. forces in the Iraq War. The *Reagan* makes it to Brisbane, Australia, where it docks to resupply and allow the crew some time on land. While at port, the ship is held hostage to an unforeseen spineless foe.

Floating idly along the coastline, translucent blue membranes absorb the warmth of shallower waters and sunny days. In heartbeat-like rhythms they pulse, taking in water and pushing it out to propel just a bit farther along the ocean current. These few blubber jellyfish going with the waves happen upon a crosscurrent pushing them farther inland and bringing with them other jellies. Soon a bloom of jellyfish is moving in undulating waves toward Brisbane's military port. It is slow yet persistent. Then one after another, after another, after tens of thousands, protean membranes push up against a metal floating fortress, where they are decidedly and insistently stopped. While bobbing against the ship's steel plates, the jellies encounter another current, one that draws vast amounts of water toward the vessel's engines, where two Westinghouse nuclear reactors are ever thirsty for the sea to cool their core. The unwitting animals have gained entry into the vital center of the warship. It is like a scene from the 1977 *Star Wars* film where rebels against the Empire discover an exhaust port that leads to the destruction of the Death Star.

One, two, tens, hundreds, and then thousands of jellyfish float inside vast tubes until, with nowhere else to go, they push and crowd and crush each other as tentacles and membranes and goo fill grates and pipes. Pounds of jellyfish flesh accumulate until it is enough to prevent water from entering the engine's condensers. The problem begins to show up on the ship's monitors as engine temperatures rise. Without the seawater, the nuclear reactors will dangerously overheat. The situation becomes worrisome, then a concern, then a threat. Officers call in the local fire department. They order systems shut down and switch vital ship organs to backup generators. So, an eight-and-a-half-billion-dollar, hundred-thousand-ton paragon of the U.S. military, with its frictionless command of the global seas, is stymied by spineless, rudderless animals who go with the flow straight into the bowels of the military vessel. These markers of animal resistance make us rethink our dominion over the globe. They have perceived, felt, and moved through the world in ways we did not foresee, since their umwelt is radically different from our own. And this difference is their opportunity to jam our cultural apparatus and machinic defenses.

Jellyfish have been around far longer than humans. They have lived on earth some seven hundred million years, making them contenders for the oldest multiorgan animal. We help the jellyfish to thrive. As we troll the seas for fish, we are hauling out their predators and their competition for food. Acidification of the oceans due to global warming helps, too. Jellies have become a worrisome problem, then a concern, and, yes, now a threat that will not stop. In oceans impoverished of fish, they multiply to fill the void. The last two decades have seen a sixfold explosion. The Sea of Japan alone has some twenty billion floating membranes of resistance. Ships wade through a sea of jellyfish and coasts are littered with their carcasses.

We did not plan for any of this. We were not trying to clear the seas to give them more room to flourish. Humans are simply going about human lives and providing for themselves and certainly not with consideration for or deference to ancient gelatinous entities. Without thinking of the implications of our actions on the worlds and ways of other beings, we have made space for the jellies. They move in soft undulations as if to the music of Debussy. But then, with more room to grow and more available foods, they increase in numbers and become thriving populations from their perspective, while from ours they have transformed into threatening hordes more aligned to Wagner's Valkyries than the poetics of Debussy.

Lisa-ann Gershwin in *Stung!* recounts a number of other destructive blooms of jellyfish, including December 10, 1999, when forty million Filipinos lost electricity. With over half the nation without power, speculation began that there was a coup or an attack by the Islamic Liberation Front on the unstable government of Joseph Estrada. The actual cause was a bloom of jellyfish that floated from the coast into the cooling system of a power plant. As the *Philippine Star* reported: "Here we are at the dawn of a new millennium, in the age of cyberspace and we are at the mercy of jellyfish." The promise of ethereal cyberspace and an enlightened new millennium gives way to a different, animal comportment on the earth. It is a coup of sorts—not overthrowing one head of state for another and not a revolution of conscious political will. Rather, the jellies are markers of durational earthly animal existence against our worlding. They reconfigure winners, losers, political stakes, and alliances.

The first recorded power outage due to jellyfish was a coal-fired plant in Australia in 1937. Their successful nuclear power plant closures include 1983, 1984, 1993, and 2011 reactors at the St. Lucie nuclear plant in Florida; the Turkey Point nuclear power plant in Florida in 1984; Torness power station in Scotland and the Shimane nuclear power plant in Japan in 2011; Diablo Canyon nuclear power plant in 2012; and Sweden's Oskarshamn nuclear power plant in 2005 and again in 2013. And the hit list is growing.

Whether it is cutting through the ocean on a battleship or feeling the nestled comfort of a home powered by the local power plant, humans are making a world with less friction and less felt connection to the rest of the earth. Culture is often an attempt to elide, evade, and forget human bodies on this earth and so dismiss contact with animals and their worlds. We use forks to avoid the messiness of food, chairs to prevent sitting on the floor and floors to avoid sitting on the earth, heating and air conditioning to ward off the weather, cars to speed our passage across landscapes and roads to help the cars, photos to capture events and books to record ideas so as not to rely on the uncertainties of memory. Almost every technology is a way of distancing us from our surroundings. Even when we get closer to nature with a zoom lens of a camera or the magnification of microscopes, it is a closeness achieved by maintaining distance. Google maps, drones, and smart bombs allow contact-free mastery of space.

Animals' frictions and animality, with their sheer corporality and insistent presences, get suppressed by culture. Occasionally, a creature goes contrary to human design and expectations. The critter becomes a clever one-off event circulated by social media, laughed off, and then quickly forgotten. But a collection of such events pops our soap bubble and causes us to rethink our relationship with the earth.

Bursting our umwelt is what philosophers Alexander R. Galloway and Eugene Thacker call "the exploit" in their book with the same title. The term comes from computer science, where hackers find a weakness in an operating system's defenses and use this opening as a portal for entry. Shifting from computers to cultural systems, we can say that the animal revolution finds holes in human cultures, which have tried to keep the frictions and messiness of the earth at a distance. Jellyfish wander into machinery designed to ease human passage across oceans. Animals exploit openings where culture has failed to sheathe the physicality of our bodies and machines.

Sitting stubbornly among our advancements is the materiality of things and the physicality of the human body itself. The revolution hacks the physicality of our technologies and our bodies. An eagle takes down a drone, a shark bites through underwater high-speed fiber-optic cables, ants infest electronics and short-circuit computers. They do not see our technologies as modes of easing human life. Rather, these objects are novel problems in their world. They respond from their soap bubbles, from the way they sense these objects. Sharks may be attuned to the electromagnetic fields strumming in currents or they may just be curious about the large eel-like rubbery black mass. Ants feel the warmth of electronics and want to snuggle up to them, while some species may detect their electromagnetics and be hypnotically drawn in. Birds and bees don't like the loud, flapping-like, animal-sized drones that seem like an annoying insect or predator. Like Uexküll, consider things from animals' perceptual apparatuses, from their points of view. In their perceptual bubbles, our technologies and machinic prosthetics are invasive beings. As we fashion our world, our objects in a shared earth reshape animal worlds as well. The animals are refashioning our fashioning to fit their own needs and wants.

Animals rub up against our attempts to ward them off, and they do so in ways that expose vulnerabilities in our systems. In a seemingly nonlinguistic but intensive mode of communicating with us, they are making us aware of a larger-than-human world. With every machine they break and human they take down, they are letting us know that there are other worlds where time, space, and perception work differently. They will not comply with the way we perceive and our way of world building. The earth is full of other creatures that do not live politely on our terms.

SEVEN
Return of the Repressed

Radioactive boar • Godzilla •
The Zone of Alienation • Killer rabbits •
An unwanted gift

Radioactive wild boar are invading towns in southern Germany. They take out a man in a wheelchair; they break through fences and roam the roads, shutting down highway traffic; they travel in packs scavenging for food. Police scramble to restore order in urban centers. The radioactive boar are armed with a postapocalyptic payload; they live in the wake of the 1986 Chernobyl nuclear disaster. By foraging on radioactive plants, the animals embody the return of a disaster many seek to repress.

Following the collapse and meltdown of a reactor at Chernobyl, more than a hundred thousand people were evacuated from the twenty-mile Exclusion Zone around the nuclear power plant. Residents exposed to the radiation suffered from radiation poisoning, leukemia, and thyroid cancer. Estimates are that some four thousand people could die from illnesses related to the accident.

Now in the Exclusion Zone, amid cracked streets overgrown with weeds, a bear paws its way across a decaying town. Markers of human habitation are slowly faltering into dilapidated ruin. Paint peels from buildings and windows have lost their glass. Signs stand askew, signaling to no one their formerly relevant information about a street name, a grocery store, café service hours. In abandoned pastures there are only sparse indications of the former crops, while native grasses convert the space into a meadow. There, short stocky horses—the only subspecies never domesticated—run wild where humans will never plant again. Thick-haired bison roam woods and fields that they have not known for centuries. Without fear of being hunted, the animals flourish in an eerily mutant, post-human wildlife sanctuary where radiation remains ten to one hundred times higher than is safe for occupancy. Rare species not seen in the region for hundreds of years have returned, including the Przewalski's horse, the European bison, the lynx, and the Eurasian brown bear.

As for the radioactive boar several hundred miles away in Germany, with an omnivorous appetite and sturdy snouts for rooting out food, they consume their landscape. They eat acorns, nuts, and insects but also unearth truffles, tubers, and mushrooms, which absorb high degrees of radioactive waste that, decades ago, drifted downwind from the power plant meltdown. In droves, the boar make their way into the nearby towns intent upon a density of food in trash cans, park bins, and alleys. Weighing in at some four hundred pounds each and with tusks and unpredictable temperaments, they are given right of way in urban areas. A coarse-haired wildness stands at odds with the orderly small-town environments in which they find themselves.

Decades hence, Chernobyl fades from memory. Generations have passed for humans. But for the radioactive elements that the disaster unleashed, life has just begun. The nuclear reactor core fire lives on, but invisibly. And the boar carry it with them. They bear the materiality of our failed technology and the indifference to life of a radioactive isotope.

Perhaps we should pay more heed to our fictions. Godzilla, a fabricated prehistoric marine reptile monster empowered by nuclear radiation, reminded Japan and the rest of the world that radioactive material is a beast more forceful and lives longer than humans can imagine. Godzilla makes the otherwise invisible nuclear threat visible. His overall indifference to humans makes him a fitting avatar for radioactive material.

The Godzilla films spawned other notable monsters, including the massive radiant moth creature Mothra, accompanied by small humanoid twins who speak on the creature's behalf. Mothra appeared in sixteen movies, including *Godzilla vs. Mothra* in 1964 and its remake in 1992 and *Rebirth of Mothra*, which, like the *Rocky* series, had a number of unfortunate sequels. Of the many Japanese monster films, *Mothra vs. Bagan* never made it past a screenplay but should have. Bagan is a massive multihorned rhino with wings, who, thousands of years ago, protected the earth from threats. Cut to the present as Bagan is released from captivity in a glacier that melts because of global warming. As protector of nature, the monster sets out to destroy humanity, which is destroying the earth. Throngs of people meet their doom while the rest plead for help. Mothra hears their cries and flies to their aid. But help is short lived as Bagan soundly savages Mothra in what would be an epic scene for an actor wearing a latex costume and a puppet moth with cardboard wings. With the monster moth defeated, all seems lost. But on a remote island, one of the moth monster's eggs hatches and a new Mothra is born. After various plot twists and suspense, the young Mothra defeats Bagan, protector of the earth. While it is clear the earth needs saving, we have a problem scripting ourselves out of existence for the betterment of the nonhuman world. It is as though *Mothra vs. Bagan* replays itself over and over again. While Bagan returns time and again, one day there may not be a Mothra spawn to save humanity.

Other nuclear disaster movies followed the Japanese franchise. In the 1954 Hollywood monster film *Them!*, an early atomic bomb test in New Mexico mutates common ants into giant, human-killing beasts. As the wise character Dr. Harold Medford (played by the *Miracle on 34th Street* Santa Edmund Gwenn) observes: "We may be witnesses to a Biblical prophecy come true: 'And there shall be destruction and darkness come upon creation and the beast shall reign over the earth.'" Mystery and ominousness ruled the day. "If these monsters got started as a result of the first atomic bomb in 1945," *Gunsmoke* cowboy actor James Arness asks of Gwenn's Medford at the film's conclusion, "what about all the others that have been exploded since then?" To which Medford replies: "Nobody knows. When Man entered the atomic age, he opened a door into a new world. What we'll eventually find in that new world, nobody can predict."

But as the proliferation of weapons gave way to the less immediately threatening use of nuclear energy for power, the peril of radiation diminished in human consciousness. It became an inhuman force under control as culture triumphed over nature. Even when disasters strike—like at Chernobyl and then later Fukushima, where radioactive boar have also been reported—humanity tends to forget.

Human control comes with a dose of repression. We bury the unwanted. We look away from the hideous progeny of our disasters. If we keep busy enough and avoid looking at minor changes, everything is fine. But the doses of repression build up, the minor starts to grow into something major. We convince ourselves that this must be a once-in-a-hundred-year sort of event. Until the once-in-a-hundred-year perfect storm seems to happen more and more often, until despite our best efforts the cancerous growth cannot be ignored.

People want to move forward from calamities like Chernobyl and Fukushima into a more hopeful, optimistic future. Our machines will carry us into brighter worlds. "We were promised flying cars!" we cry. We who feel a technocultural imperative want to forget our vulnerability as bodies on earth and get on with our cultural lives. There are quips to tweet, dinners to serve, and a veneer of stability and progress to maintain. But recall the exploit—that humans and animals may live in different perceptual worlds but corporeally share the same earth—which the revolution leverages as an opening to rupture culture. The animals won't let us forget our disasters or the earth we share. They carry our past along with them. In eastern Germany, the radioactive cesium-137 levels in wild boar are six times the European Union limits for safe hunting and consumption of game. Geiger counter stations stand sentinel to remind citizens of an invisible toxicity. Hunters can haul their game to check for radiation levels and the machines read the toxicity in flesh and fur. There is nowhere to run from the geological time of radiation and the evolutionary time of animals carrying the disaster's ongoing effects back to us. They are the return of the repressed!

These unpaid actors of ecological remembering are not tricked out in Japanese monster costumes. There is no man in a latex suit, no puppetry, no scale models. The Exclusion Zone and sanctuary around Chernobyl is also known by the almost existential title "Zone of Alienation." Who is alienated if not we humans? First from a time outside of human time—the half-life of radioactive elements—and then from physical bodies that do not conform to planned technological progress. Even though these beasts seem more modest than our fictions imagined, creatures like the boar have become the real Godzillas and Bagans, invading our cities tusk and snout to remind us of the (breached) boundaries of human control.

Americans are not immune to the invasions found in Europe and Asia. In November 2010, a small brown rabbit nibbles his way to the border of the Hanford nuclear site in Washington state—the largest nuclear site in the Western Hemisphere. There in a grassy plain are some tasty morsels in a small enclosure. Tentatively, with whiskers twitching, he enters the opening and *snap*, the rabbit is trapped in a cage. Hopping and pushing frantically against the shut metal door, the coney looks to make an escape. After what seems like hours, the animal gives up and awaits his fate. Humans in white suits come and pick up the box. The rabbit, now weary and wary, is hauled from the confines and "inspections" begin. Later, a report emerges from the lab. The animal is highly contaminated with radioactive cesium.

Hanford is the site of the first nuclear reactor and the facility that fed plutonium to the "Fat Man" bomb dropped on Nagasaki, Japan. The bunny seems innocuous enough, until one realizes that it, being a rabbit, breeds. There must be more out there inadvertently foraging for radiation from the site and adding more potential carriers of radioactivity. And why stop at rabbits, since there are myriad animals across the Hanford site? How many? As the movie character Dr. Medford says, nobody knows.

The Hanford reactor was put out of service in 1988 but left behind millions of tons of solid waste and hundreds of billions of gallons of liquid waste from its decades of producing plutonium. The radioactive material is buried underground in dark pits and holding ponds where—like the repression of a bad memory—it has been forgotten. As the U.S. Department of Energy explains: "Depending on when the waste was buried, records about what was buried and where it was buried can be either very good, or in some cases, very bad." As inhuman time moves onward, liquid waste continues to soak into the soil. The membranes designed to separate nature and culture have worn down, and radioactive rabbits are the result. The bad memories return.

A decade after Fukushima, three decades after Chernobyl, and seven after the Manhattan Project, the boar and bunnies wandering in the wake of disaster continue to bring us a gift. It is the same lesson we began to explore with Godzilla but quickly relegated only to fiction: the human trajectory of technological and social progress has produced by-products that linger on a scale of time and space far larger than humanity can easily understand. The Chernobyl boar are not just visitors from the past, it turns out. Thanks to the longevity of radiation, they are also visitors from the future, a future of continued radioactivity and leaking between the membranes of culture and nature. To take their gift seriously would require accepting the repressed detritus of human progress and incorporating that aftermath into the idea of progress rather than believing that it remains safely buried, cordoned off, and forgotten. Are we willing to accept such a gift?

It would mean looking steadfastly at a lengthy catalog of cultures' mishaps written across the earth and in the bodies of animals. We would not be able to script a Mothra to save us. And we may have to abandon all hope of flying cars and other optimistic technocapitalist promises at the expense of other life on earth. Instead, it would be a hospitality to what revolutionary animals are telling us, bringing to us time and again. It would be fashioning our technologies to accommodate them and their messages. World-famous biologist E. O. Wilson has proposed a half-earth plan. We live in cities and make use of strategically planned areas over half of the earth and leave the rest for all other beings. Perhaps it seems impossible with the messy intermingling of humans and animals, but Wilson is opening himself to the message brought by the return of the repressed. He has created a speculative proposition and is asking us to make room for animals other than ourselves.

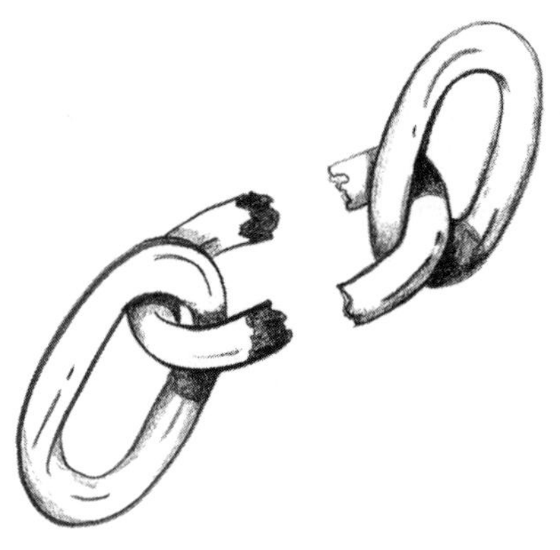

INTERLUDE
Multiplicities

A flamboyance of flamingos
A swarm of bees
A bloat of hippopotamuses
A convocation of eagles
A colony of rabbits
An unkindness of ravens
An obstinacy of buffalo
A smack of jellyfish
A mob of kangaroos
A troop of kangaroos
A court of kangaroos
A shrewdness of apes
A zeal of zebras
A leap of leopards
A sledge of cranes
A cete of badgers
A shadow of jaguars
A labor of moles
A team of pigs
A caravan of camels
A colony of goats
A quiver of cobras
A destruction of wild cats
A bask of crocodiles
A lounge of lizards
A pack of wolves
A gang of elk
A cast of falcons
A skulk of foxes

A deceit of lapwings
A business of ferrets
An army of frogs
A tower of giraffes
A band of gorillas
A cackle of hyenas
A conspiracy of lemurs
A prickle of porcupines
A colony of rats
A shiver of sharks
A stench of skunks
A knot of toads
A fever of stingrays
An ambush of tigers
A murder of crows
A wake of buzzards

PART II

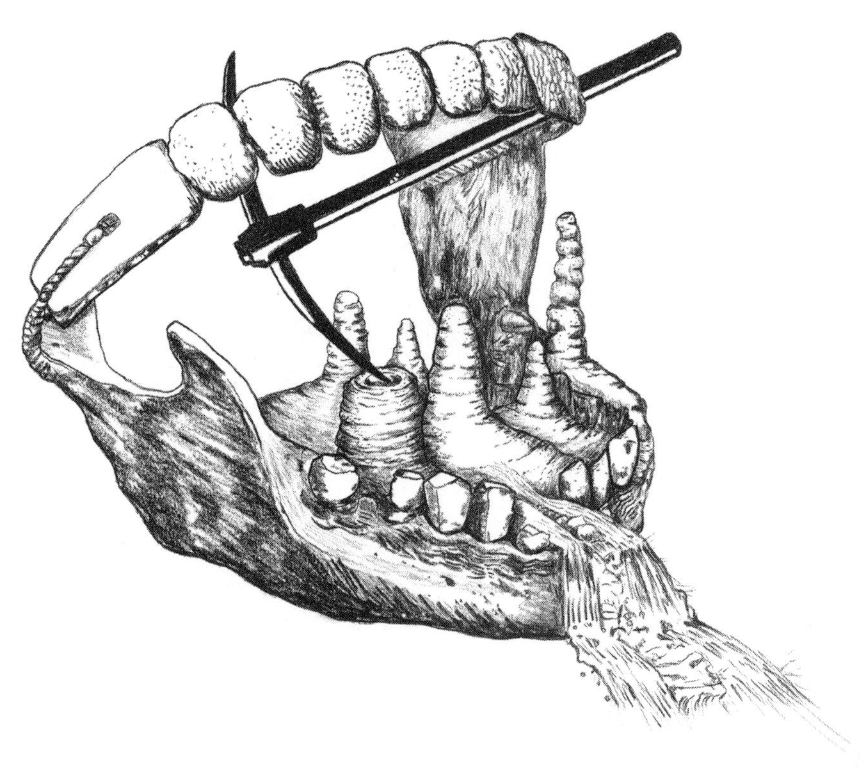

EIGHT
The State of the Union

Multispecies teeth • Vacancy and mortality • A talisman

The first president of the United States was a cyborg. George Washington's various dentures were a cabinet of curiosity holding eighteenth-century natural-history lessons, economic exchanges, and political relations. Contrary to popular belief, none of his teeth were made of wood. Instead, his dentures were a collection of bones and metals. He bought teeth from some of his slaves, who certainly felt the asymmetrical power of the exchange. Other teeth likely were from corpses. Some were from cows and horses fashioned to fit a human mouth. Teeth were made from elephant tusks. Perhaps some were from whalebone. Unsubstantiated rumor is that bone from the first unearthed mastodon was sculpted into a tooth and given as a gift to the president. Found in the verdant Hudson Valley, the fifteen-thousand-year-old mastodon remains were hailed as national grandeur over and against European speculation that America had only diminutive animals and, by symbolic extension, that the American people were simple and small-minded. Washington's teeth would prove Europe wrong. Also in his mouth were parts of the earth itself, mined and refined metals including tin, copper, and brass alloy.

Rolling his tongue against these teeth, Washington could feel prehistorical times. He would clench the tusk of an animal from a distant continent against bone from a creature of the deep seas. While sipping a brandy, he might imagine cold, damp mines, catch a hint of the fire and smoke from smelting and the acrid taste of minerals. He might muse on the relationship between cow teeth and the beef he was eating or his horse teeth and the bit and bridle he held like a connective string to the teeth of his warhorse, Nelson. Did he ever look at one of his servants, look into his mouth and the vacuous space among the teeth, and, while holding the slave's tooth in his own mouth, think of the relationship between the two of them beyond a commodity exchange?

In the dark of evening, after company has left and in the quiet of his home, Washington unburdens himself of the crafted yet still uncomfortably fitted and heavy form of bone and metal. His jaw unclenches and he feels the absence of shape in his mouth. Before cleaning it and putting the apparatus away, he looks at it glimmering with saliva in candlelight. Now neither in his mouth nor in the humans and animals and earth whence the dentures came, he thinks of those absences, the incompleteness that he and these others have. He holds their missing jigsaw puzzle pieces in his hand. In the dark vacuity of the night, the dark emptiness of his mouth, and out there the emptiness of a space that once held substance, he feels a connecting mortality. He is aware that his time on earth is bought at the expense of others, but still his arc will lead to the grave. He and the nation he leads are propped up by what is harvested. In the waywardness of the night, the teeth feel like a haunted talisman that holds the specters of other lives.

How much of these other lives resonate in the talisman that the president carries in his cavernous mouth? The animal revolution has hacked into the head of the head of state by exploiting the decay of his teeth. The words he speaks come from a mouth filled by slaves, corpses, and animals. His words vibrate from elements of deep time and earth. And his food is masticated by them as well. This towering leader might make use of humans, animals, and minerals, but what a figure for a future power as the revolution plants itself intimately within the president. Washington has felt and daily knows that his voice is not his own. If the president is the mouthpiece for the nation, the country speaks as a heterogenous, more-than-human, mineral-animal-human state of the union. It is a place of many beings with many lives, each in their worlds, grinding against each other, and negotiating a shared earth.

NINE

Exploit Continued, or The Exploit Exploits

A duck mole • The Trojan horse and a few rabbits • Killer rabbits • Poison toads • Mussels and the mighty Colorado • *I Am Legend*

The eighteenth century was an age of British exploration and conquest, with the finding and naming of things as if laying claim to knowing them. Oak-mast vessels fortified with crew, provisions, and the latest scientific equipment creak across the high seas in search of new lands, peoples, flora, and fauna. There are legends of a large undiscovered landmass in the Pacific. With crudely drawn maps and then incrementally more accurate ones, cutting and looping across large swaths of the ocean, British explorers eventually happen upon a land they only half speculated might exist—Australia. The wildlife is like nothing Europeans had ever seen. They figure the kangaroo to be some sort of dog, and as for the wombat, it crosses species classifications and stumps men of learning. When ships return from the remote continent with a taxidermied platypus, British natural historians think it is a joke. With webbed feet and a bill like a duck but fur like a mole, surely this "duck mole" must be a prank animal like the famed taxidermy mermaids or today's mythical jackalope.

These strange animals are a story first composed in deep time, when tectonic rumbling caused the island that is now

Australia to break away from the rest of the continents. As a result, life evolved differently than elsewhere. Monotremes such as the platypus and echidna are examples of the unique biodiversity of the continent. Their evolutionary differences are the reason early European explorers were dumbfounded by such wildlife. It is like stepping onto another planet. In Australia, survival of the fittest means a fit to a different ecology; it means growing a duck-like bill or having a pouch to rear a young marsupial. But these wonderful animals' fittedness to their environment is endangered by the strange humans and their cargo sailing from very different ecological landscapes.

Under the decks of the teeming British vessels await furred and feathered foreigners biding their time to get back on land. These beasts establish a colonial stronghold. The arrival of species never before encountered on the continent creates unforeseen consequences. They sniff, paw, thump, perch, leer, and learn these new domains. The European settlers have unwittingly become the exploit, the Trojan horse, by which novel life is introduced to Australia. Now with a clawhold into new lands, the interlopers run rampant over a terrain not evolved to defend itself against these colonizers.

The First Fleet sent by King George to establish a presence in Australia set loose a colony of European rabbits. Later, colonist Thomas Austin proclaimed, "The introduction of a few rabbits could do little harm and might provide a touch of home, in addition to a spot of hunting." But rabbits were never really under easy control in Europe, and even less so abroad. Austin's twenty-four rabbits have become two hundred million. They munch down native plants, cause soil erosion, and generally ravage the landscape. They have pushed out native species, causing the decline of the greater bilby, yellow-footed rock-wallaby, southern and northern hairy-nosed wombats, the malleefowl, and the plains-wanderer. In short, bioinvasive bunnies are causing ecological mayhem.

Then things get serious. In an exploit of the exploit. They are there not only to take over the terrain of other animals but also to wreak havoc on weaknesses within human systems. With a brilliant lack of discrimination between wild and cultivated plants, rabbits devastate not only the grasslands but also agricultural fields. As they nibble their way westward across the continent, settlers take to cursing Thomas Austin and start plotting defensive measures. Perhaps the most ambitious plan was the 1907 Rabbit Proof Fence, later fortified by giving it the more intimidating sounding name, the State Barrier Fence of Western Australia. Taking six years to complete, spanning 1,139 miles with more than half a million fence posts, the three-and-a-half-foot-high wire barrier has become a strange and desperate measure. Colonizing Brits brought the animals over for sport only to find themselves the portal for carrying rabbits to more fertile ground. Sport has now become an all-out war. *Star Trek*'s "Trouble with Tribbles" seems an apt analogy.

But rabbits are not only hopping. The European rabbit in particular is known for creating networks of burrows and warrens. Consequently, there were bound to be problems. Following a natural disposition, small paws claw the dirt. They dig down and then under the fences. And with so much fence, inevitable holes, tears, or punctures provide rabbit-size openings in what becomes a porous barrier. To repair and secure the fence, teams driving camels and now ATVs patrol the full thousand-plus miles. Rapid-fire artillery is deployed to fend off the fluffy vermin. Despite all this, colonies of rabbits have made it to the promised lands of Western Australia and its agricultural fields. To this day, the fence mitigates but does not stop rabbits, dingoes, and other unwanted beasts.

The reason *Monty Python and the Holy Grail* includes killer rabbits is that even in medieval Europe the animals were just beyond the control of humans. Bibles and bestiaries are decorated with colorful illustrations of strange animal activity, including rabbits cutting the heads off of knights, clubbing humans, and jousting with them. Among the illustrated havoc is a bewildering image found in the Yates Thompson 8 Breviary of a very determined rabbit riding a human-headed snail into a jousting match with a dog. To the scribes, rabbits stood for unlicensed sexual reproduction and earth religions that lingered in Europe as cults against Christianity and its policing of sexual morals. Now European rabbits, actual rabbits, in Australia activate their sexual proclivity and multiply in a joust against the human ordering of things.

Rabbits are only the beginning of the hopping menace.

Yet again, the problems begin with British colonies on the continent. Amid the ark-like array of novelties brought from afar, the First Fleet introduces sugarcane in the hopes of creating profitable plantations as elsewhere across the globe. But this means trying to find the right terroir, preparing the soil, and entrapping a workforce for the backbreaking labors. By the mid-nineteenth century, cane translates into money and eventually becomes a major export for British Australia. But amid the spread of sugarcane, nature bites back. The native greyback cane beetle and the Frenchi cane beetle develop a sweet tooth. They infest sugarcane fields and eat away the profits.

With the nation's major export at risk, people start paying attention. White-collared executives gather in boardrooms and call in scientists and farmers for field reports. In 1935, the Bureau of Sugar Experiment Stations unleashes its solution. A ship arrives bearing three thousand cane toads from the Americas. It is an innocuous-enough-looking amphibian. With brown skin and dark brown spots, squatting on the ground, it looks like a textbook image of a toad. What could go wrong? The toad will eat the Aussie beetles and the problems will be solved. Or so went the boardroom drawings of the plan.

The ability to eat beetles is a mere Trojan horse, a gift to humans that allowed the toads entry into new terrains. The natural predators of cane toads were thousands of miles away. And despite its innocent, even lethargic, toad look, the cane toad is highly poisonous. Because of this poison, not all seemingly natural killers of toads can digest them and live to tell about it. Without select species of alligators, ibis, various species of fish, snakes, and eels evolutionarily adapted to prey upon the cane toad, these squat amphibians are free to roam without fear of retribution. Over the next eighty years, the toads multiply to over two hundred million today, thus rivaling the rabbit population.

While the sugarcane industry wants to use them as a biological, bioinvasive weapon, the toads will exploit what Aussies believe is a gift. Eating pests is just a way to hack into the agricultural and financial systems and take over. The toads will exploit being exploited. Like the biblical plagues of Egypt, cane toads infest Australia. "They will come up into your house and your bedroom and onto your bed, into the houses of your officials and on your people," warns Exodus. Now the innocuous-looking reptile unveils itself to be an amphibian menace.

In knots, toads sun themselves on major roadways. Enclaves inhabit parks and public thoroughfares. In some areas, a home is not complete without its cane toad problem. Golfers with a sadistic malice have taken to practicing their swings on their rotund bodies. But the brilliance of these animals is that they need not be cunning or swift. They simply have to multiply at amazing rates producing a further mass of cane toads, who will exponentially continue to reproduce. The bioinvasives' numbers press up against our technological means of managing them. We call them stupid, yet we are the ones who introduced them to new lands, and their so-called intellectual stupidity comes with a smart corporality that outmaneuvers our ability to decrease their population.

The triumph of the toads reminds us of animal worlds and an ecological earth beyond our conceptual grasp. Humans have been far too reductive in system analysis of problems and solutions. Sugarcane does not simply produce a pleasing product for human consumption, and it is more than a sweet, refined, granular whiteness as economic commodity. It is also a food for beetles who don't invest in our economies or semiotics. And cane toads are not a contained unit solution to a cane beetle problem (a problem that is only a problem for us but not for the beetles). Instead, toads have worlds and wants of their own. Unleashing the cane toad as a solution also means opening an amphibian Pandora's box of unforeseeable toad proclivities tucked away in their corpulent mass.

They leverage the space where cultural systems—in this case, agricultural and financial—meet the physical realities of an earth shared by humans and toads. Humans need this shared space and want to occupy it with farms and profits, with real estate and a national flag. But the toads living in their perceptual worlds take up the shared space in ways we haven't anticipated. This blind spot becomes our weakness and their triumph.

The tale of bioinvasive animals repeats itself across time and space. They come from another world far, far away and are introduced as benign forces for good. They land in a place unprepared for their arrival. They thrive and then overload the environment. And then they overwhelm our capacities to manage them as they clog human systems. But a second version of this tale is even more common. They do not come as a force for good but as stowaways. They are the unintended consequence of our travel.

One summer sometime in the 1980s, a cargo vessel somewhere along the Dnieper River starts to pump water into its ballast to even its load of freight destined for the West. Upended in the suction, a bed of small striped mussels bumps along hoses and pipes only to find themselves in the watery bilge of an international cargo ship. After weeks of a dark, wet, sloshing journey, arriving thousands of miles away in Lake Erie, the mussels are drawn into the propelling water discharged from the ship. The migrant mollusks have found a new home. Being less than an inch each, who is to notice their arrival? Problems only come later. By the early 1990s, they are discovered to be disturbing the plankton ecosystem of Lake Michigan. Eventually, the invaders displace about half of the native bivalve species of the Midwest.

In January 2007, the zebra mussels are found much farther west, in the man-made reservoirs of the Colorado River. The mussels cling to surfaces where they can take in large amounts of water to capture plankton and microorganisms. The river's dams, built to supply energy and water to tens of millions of Americans in the Southwest, are an excellent habitat. Mussels pile one upon the other in groups and clumps until the inches become feet of shells on the dams' turbines. These tiny mollusks clog the hundreds of tons of concrete engineering along the mighty Colorado. The epidemic is such that dams are compelled to shut down turbines and perform regular maintenance to rid themselves of the encrusted biomass.

What is happening here? We don't take account of the ways animals occupy the earth and how they perceive the world. When they do push back in one incident or another, we laugh it off or we repress the implications. Culture is not hospitable enough to allow space for animal umwelts. We don't anticipate the multiplication of animals in swarms, mobs, troops, teams, colonies, and gangs. We can smile and pity one animal fighting against our mighty systems, and we can repress one-off vectors of nonhuman destruction. But the threats build, becoming larger, multiplying faster than our abilities to contain them. We give these animals the name "pests" or "bioinvasives" in the hope that the words will assure us we know what we are doing. Cast like a spell, rightly naming gives hopes of containment . . . until we realize that casting does not dispel the problem but seems to summon the inhumanity that pushes against us.

Of the thousands upon thousands of invasive species and millions upon millions of actual critters doing the invading, it is the human animal that stands out as the most intrusive and destructive. From cities to farmlands to extractive mines and human-made pollutants, we weigh upon the earth. Maybe, just maybe, they are not exploiting our systems—maybe this is payback for us exploiting theirs.

Consider our life-sustaining practice of farming. Weed killers also kill the vibrant microbes in topsoil; we replace lost nutrients with fertilizers that then, during industrial waterings and natural rains, create runoff into riverways that spill into oceans, making for miles of dead zones uninhabitable by aquatic life. Oceans are also dumping grounds for plastics that create the Great Pacific Garbage Patch, the size of Texas, set afloat on the high seas. Cargo ships from the Americas to Asia have to plot courses around the petrochemical monstrosity. Smaller microplastics from shampoos, dermatology creams, and even clothing find their way into the water, where they disturb the digestive systems of small marine animals. There are offshore oil drilling "spills," such as Deepwater Horizon in the Gulf of Mexico, which killed tens of thousands of birds, marine mammals, and fish. Extraction of coal, copper, uranium, and a range of precious rare earth elements all add to the destruction by contributing miles of barren earth from mountaintop removal and open-pit mining. With global warming, Greenland is changing and the Arctic shrinking. In Siberia, melting ice has released long-buried anthrax. The Coca-Cola polar bear from the 1920s is no longer a chipper novelty but rather a melancholy icon of loss. Less charismatic species go extinct at incomprehensible rates. Humans and our technologies are a Darwinian vector forcing animals to adapt to rapidly changing environments or die. Our situation is much like that of Robert Neville, the main character from the book and film *I Am Legend*. In the fictional postapocalyptic future, most humans have become vampires (or, in the film, zombie vampires) because of an unexplained pandemic.

Neville barricades himself from the infected humans, who, in traditional vampire fashion, only come out at night and will die in direct sunlight. He hunts down and kills many of their horde and experiments on others as he tries to find a way to cure them. At the end of the book and film, Neville is trapped and captured. Reflecting on the destruction he has wrought upon this infected breed of humans, he realizes that they are not the monsters—he is! He is not saving anyone. Instead, from their point of view, Neville is the destructive Other, the creature about which they write legends. And today humans are the bioinvasive monsters destroying local and global ecosystems. Is it any wonder the animals push back against human monstrosity?

TEN
Other Intelligences

The Inheritors • Why there is no intelligent life • Chimps and bonobos • An ark within us

A year after William Golding's dark tale of innate human violence in *Lord of the Flies,* he published a little-known novel, *The Inheritors.* The story is written from the point of view of several members of a small clan some thirty thousand years ago. They go about their daily rituals of gathering berries, seeds, and grasses. They haul water and struggle to keep warm. They play games and chase each other for sport. When one of them dies suddenly, the others bury her in a plain yet moving ritual while each member of the clan tries to come to terms with the loss, the once living now dead. Golding writes in plain words and sentences. The characters' thoughts are simple but striving for something just beyond their grasp. It is as if we are witnessing the dawn of self-reflective consciousness with a dreamy haze and occasional clarity. They are in a drowsy awakening.

The characters know themselves as the People. Later in the novel, a foreign group appears, whom they refer to as the New People. Golding describes one of the People reacting to the new clan moving into their terrain. He wants to see them as People but realizes that they are not like the people whose names he knows: "Lok felt the shock of a man who has trusted to a bough that is not there. He understood in a kind of upside-down sensation that there was no Mal face, Fa face, Lok face concealed under the bone. It was skin." Slowly, as the novel unfolds, the reader begins to realize with the same shock as Lok that the story is written from the point of view of the Neanderthal and that the New People are our ancestors, us humans. The New People carry strange sticks that fly fast through the air. Lok's senses tingle. Something is different here. He thinks they are trying to send him a gift. Only later does he realize that they are not friends and this is not play. He is being hunted; they want to hurt his flesh. The darkness of *Lord of the Flies* begins to creep into the novel. Neanderthal openness confronts human aggression.

In occasional smugness or in despair, we may ask ourselves why there aren't other intelligent creatures on this planet, ones with language and art and something like what we call higher-order thinking. One answer to this question is simple: it is because we killed them. Humans have leaned into the world with a kind of innate violence that perceives any other intelligent beings as a threat, competitors for food, intruders on ground that should be ours.

Neuroscientists have looked into the deep history of evolution to find the origins for human attitudinal comportment. They use computer parsing of genetic material buried in the cells of human and animal remains. From the shiny digital displays of ancient material, they begin to weave a story. Of the many clever species on this planet, our closest living relatives are chimps, and it turns out they display active aggression similar to humans. Such was the story until about a decade ago, when scientists sequenced the genome of bonobos. We share 99 percent of our DNA with these smaller chimpanzees (*Pan paniscus*, or the "diminutive Pan"), which is the same percentage we share with chimps (*Pan troglodytes*). Deciphering the story from genomes, humans split from the ur-ancestor of bonobos and chimps some eight million years ago, and bonobos and chimps split from each other less than a million years ago.

There are many evolutionary paths, some leading to dead ends and trait extinction, some sufficiently robust to spawn new branching paths. Resources are mustered, evolutionary traits unfold, bets are placed. Among all the crossroads of selection, some affect species' attitudinal comportment. So, while chimps and humans have similar aggression, our other *Pan* ancestor does not. As ethnologist Frans de Waal, director of Living Links at the Yerkes National Primate Research Center, explains in "Bonobo Sex and Society": "The species is best characterized as female-centered and egalitarian and as one that substitutes sex for aggression. Whereas in most other species sexual behavior is a fairly distinct category, in the bonobo it is part and parcel of social relations." Conflicts in a group are resolved through sex rather than fighting, and aggression gets soothed by an embrace or affectionate touch.

With equal genomic similarity to chimps and bonobos, it seems most human societies have tipped toward aggression rather than sex to diffuse tension. Perhaps the disposition is genomic, perhaps it is cultural. By contrast, there is the Neanderthal. Humans split from them roughly half a million years ago, give or take a couple hundred thousand years. From ethnographic and anthropological speculation, Neanderthals were a kinder species, perhaps more like the bonobo. One lesson from Golding's *The Inheritors* is that human violence ended a gentler species with its way of thinking and being in the world.

But this end is not exactly an end. Whether through affection or malevolence, humans and Neanderthals had sex and produced offspring. Some 10 percent of genetic material in modern Europeans is Neanderthal. So, while humans can read Golding's story as a victory through violence, the trace of those encounters lives on. We have inherited the earth, but also some humans have inherited and carry the material evidence of struggles and passions from some thirty thousand years ago. Here in human bodies is the exploit from another race. While their culture is lost in time, they have made their way into the genetic material of bodies passed down for generations upon generations. Like an ark, humans bear the others within us. Is it possible that nonhuman comportment and thinking lingers there too?

This specter from another time and another people can become a way out of the self-enclosed human island with its *Lord of the Flies* dispositions. While genetic dispositions are not the only factor in human behavior, they are both a factor and a figure for ways we can live on this earth.

Make Love, Not War could be more than a 1960s counterculture slogan. Within humans are alternative modes and dispositions that can be activated to voice a different world not only for ourselves but for the other animals.

ELEVEN
Giving Voice

A golden record in space • The necessary alien • Wittgenstein's lion problem • Big Ear • The open secret • Lovecraft, Poe, and "Tekeli-li!" • William Shatner

In the expansive darkness of deep space float two tiny metal objects. Each has a giant radar eye in its center and long, spindly insect antennae extending from its golden decagon body. With a purposeful drift, they slip farther from us. The crafts shimmer alone in a cold, sunless beyond, moving past our solar system and into the boundary where solar winds get pushed back by the winds from other suns. Over the past forty years, the Voyager spacecrafts have seen many lives and many worlds. Long ago they outlived their primary purpose of photographing Jupiter and Saturn. The images they beamed back to Earth made the covers of *Time* and *National Geographic* in their full-color, glossy glory.

The Voyagers are now on their long secondary mission to extend human technology and exploration into the dark unknown. Among the twinkle of thousands upon thousands of stars, they venture beyond the lifetimes of any foreseeable human generations. Forty thousand years from now the Voyagers will come within two light-years of star AC +79 3888 (seventeen and a half light-years from Earth). The 1970s metal, wire, and gears floating in space is in itself a message to the cosmos that we exist, we make things, we send these things into space. Not only are the crafts a message, they also carry a message on them. Each vessel has a golden metal disc with gnostic carvings of circles, vectors, and waves explaining how the disc is to be activated. The discs are phonographic records that contain in their grooves sounds from our small, blue marble orbiting a minor star.

If the discs are played correctly, one can hear baritone frogs and laughing hyenas, the murmurings of water, and the howl of wind accompanied by a complementary howl of wild dogs. Then, in a linear acoustic chronology, civilization emerges. First there are footsteps and a haunting laughter, followed by what sounds like static but is intended as fire. Human speech, hammers and saws, tractors, Morse code, trains, and a rocket. War and the atomic bomb are left out, of course, and animal voices recede with the exception of an occasional baa of a domesticated sheep or neigh of a horse. There are spoken greetings in fifty-five languages and world music from different cultures, including some Western classical—the strings of Mozart, the piano of Beethoven, and early rock guitar as Chuck Berry belts out "Johnny B. Goode." The Beatles' "Here Comes the Sun" was supposed to be on the record but while the band liked the idea, they did not own the copyright and NASA could not secure the needed legal permissions.

Let us suppose an alien found this disc. Suppose again it figured a way to unlock its inscribed contents. What is this alien to make of such a *wunderkammer* of sound? Three decades after the launch of the space probes, a SETI scientist went rogue with the website Scrambles of Earth; the site claims that, yes, the Voyager music was found by aliens, but with the punchline that they remixed it to their liking and beamed it back to Earth, explaining: "Aliens want state-of-art mega-hip scene. But music on Voyager record is hundreds of years old! No synthesizer, no drum machine, no dance remix! No wondering that, in ten year, not one alien has called! Aliens hear old tunage, puke out all kidneys from ear-pain. Say: 'Gaia beings are coma-toast lamo Gilligans.'" Scrambles' parodic "alien" music is a haunting techno remix that makes it clear that no alien would understand the Voyager records and the selection overseen by cosmologist Carl Sagan. It turns out that the golden discs billions of miles from Earth are for us.

The tongue in (alien) cheek of Scrambles fits what cultural theorist Richard Doyle considers the truly alien: "The very essence of the alien presence is its characteristic ability to proliferate and mutate, disturbing the various taxonomical categories that we bring to bear on 'them.'" One UFO investigator summarizes it this way: we "are meant to be baffled." Alien life is fundamentally, well, alien. Why would we think they know baritone frog sounds or that a crunching sound is human footsteps on pebbles or a static crackling is fire? Conversely, what are they like and how on earth could we ever communicate with them?

This is not a new question. In the eighteenth century, philosopher Immanuel Kant thought a lot about what it means to be a rational being. It was an era of European exploration. Ships came back with tales of strange customs of foreign peoples across the globe. Europeans were discovering for themselves that there are humans in an array of shapes and sizes and skin colors with different clothing, music, dances, and stories.

Reading and witnessing the wonders of other worlds, Kant ponders what makes us all human and what holds us all together despite differences. Putting pen to paper he scratches out propositions and follows these to deductions and conclusions, then circles around again, crossing out ideas, adding new ones, taking up other authors' arguments with praise or corrections. He lectures on the topic to try out new ideas. Then, from the comfort and safety of his modest home in what was then Königsberg, Germany, he publishes his conclusions on human societies as *Anthropology from a Pragmatic Point of View.*

Kant makes the Eurocentric observation that all races and civilizations are less rational than Europeans. It is an obviously self-serving move and rather common in European natural history of the time. But then he pushes beyond his peers: how do we know humans are rational creatures? We would have to have another species of rational beings for comparison. Or, as he says, "The highest concept of species may be that of a terrestrial rational being, but we will not be able to describe its characteristics because we do not know of a nonterrestrial rational being which would enable us to refer to its properties and consequently classify that terrestrial being as rational." As philosopher David Clark explains, Kant is proposing alien intelligence; in fact, he is staking our own rational intelligence on the necessity of nonterrestrial rational beings. Otherwise said, we need smart aliens. We need them because they will confirm our intelligence. And now the decades of Hollywood alien movies make sense. Be they killer aliens in *Independence Day* or the cute alien-in-a-bicycle-basket *E.T.* sort, the films are confirming that there is intelligent life by which to measure our own—even when we measure up poorly.

In Spielberg's *Close Encounters of the Third Kind,* a stereotypical model of a nonterrestrial spaceship lands in the wilds of Wyoming. With the ominous geological protrusion of Devils Tower for a backdrop, the ship emits lights and sounds and eventually a five-tone pattern for which the film became known: d e c C G. Re mi do do so. Apparently, the aliens (with the help of composer John Williams) are able to communicate in a twelve-tone chromatic scale. From Scrambles of Earth to *Close Encounters,* tunage opens an avenue for conversation without human speech and allows us to think about an inhuman resonance between beings. What are the limits and possibilities of such a transbeing vocality? If we can't understand them, can we at least give voice and hear the voice of baffling beings? Could we reside together without understanding one another?

Listen, those who have ears to hear—as the biblical passage

goes. Listen and be transported, carried—or, dare I say, abducted—by a voice out there now transmitted through waves into your brain. Listening and recognizing voice, we give voice in return. Voice gives. It opens the body through mouth and lungs or whatever alien apparatus opens onto the world. Voice creates resonance from within and gets projected outward. It is each creature's signature of unique opening. It is being vulnerable by saying "I am here" with this body propelling air and soundwaves outward into space.

If the exploit allows animals to use the materiality of culture and the physicality of human bodies to gain access to our systems, then, conversely, our physicality is an opportunity to reach out to others—aliens and animals alike—with whom we share the Earth or beyond Earth to the cosmos. Voice marks that space between, that third space, a disjunctive synthesis that is both the connection between beings and the distance and difference among them. The space between—like a chasm between aliens, animals, and humans—seems daunting and deep and with no way to cross. Yet into it we voice our joys and desperations for connectedness, however anthropocentric or ineffectual. They lose their tether to the speaker and echo, become distorted, words unravel into sounds, the meaning of which is solely the sounds themselves and, caught in some nook or deep into a ravine, they are lost to us. And still more sounds arrive in this space between humans or from those others across the abyss. The sounds clash, stack, entwine. They begin to fill the place where words are lost. They begin to make an ephemeral bridge that one dare not walk across but sees like a specter, a vision, a hallucination, a dream, a want, a despair, and a hope, until the seeing becomes a tapestry upon the rising and collapsing traversal architecture. The fabric waves amid the winds of sound while its indecipherable image becomes tattered and loses threads.

The architectures of sound are here among us. It is possible that aliens are signaling to us, but we have no way of understanding. In the 1960s, in preparation to understand potential encounters with aliens, Carl Sagan and dolphin researcher John C. Lilly postulated that we should first look at animals here who are the aliens among us. We need not look into deep space but, dropping our gaze, we will find that intelligent aliens inhabit this world. But how are we to understand them? It is a problem of connecting their speech to meaning. In the early twentieth century, Austrian philosopher Ludwig Wittgenstein famously noted in *Philosphical Investigations* that the problem of communication is not just about words but how the user and translator think similarly or differently about what the words mean. Wittgenstein summarizes it this way: "If a lion could talk, we could not understand him." To have a conversation between beings, their words must fasten to the same meaning. As Wittgenstein says: "To imagine a language is to imagine a form of life." A lion's sense of life in the world is different; so its words, the meaning of the words, and its use of language would be different—even, as Wittgenstein says—unintelligible, baffling, and, yes, alien.

Anthropologist Kathryn Denning uses the example of the arctic snowy owl. Observers have noticed for generations that the white-feathered owl blending in with the arctic landscape occasionally turns its chest toward the sun. It was colloquially thought they must be sunning themselves. Only recently have scientists realized that male snowy owls do this to reflect the sunlight off their feathers as a warning sign to other owls. The signals have been there all along, like an open secret waiting to be understood. Or consider the tarsiers. For decades, scientists watched these tiny, large-eyed, long-tailed primates. Occasionally, the animals would open and close their mouths but no audible sound came out. Perhaps, they were cooling themselves off, much like a dog panting. Then in the last decade, digitized field recordings showed that they are sending and receiving ultrasonic calls from one another. Tarsiers are one of the few mammals with this acoustic ability. While our cone of awareness to other modes of communication is slowly widening, we are left to wonder what signs from animals and aliens are in the open but closed to our comprehension.

Dutifully cleaning out the chimney to his eighteenth-century manor house outside of London, David Martin finds the skeletal remains of a pigeon. Attached to the bird is a small, red capsule containing a rather oblique set of handwritten capital letters on a scrap of paper with letterhead reading "Pigeon Service" and addressed to a mysterious XOõ. British military cryptographers surmise it is a message from WWII and likely from the French front. Despite best efforts, they could not break the code. The message must be written in a random set of letters whose deciphering would be known only to the sender and receiver. We can see the note, read the letters, but while out in the open, the meaning of the signs remains opaque to us.

How many other signs are there in the open but indecipherable because we do not know or understand the sender or receiver, or even know that these are signs? And what of the messages lost in time, such as the coo roo-c'too-coo of the passenger pigeon, the yips and howls of a Tasmanian tiger, or the clicks and whines of the baiji white dolphin, all of whom are now extinct?

We might not have ears adapted to the frequencies of alien voices that are there in the space between us like an open secret. Nevertheless, we continue to listen. In 1977, Ohio State University's Big Ear radio telescope picked up a signal emanating from deep space and transmitted for seventy-two seconds at a frequency of 1420 MHz. Hydrogen, which is the most common element in the universe, resonates near 1420 MHz. Aliens wishing to be heard might broadcast at this frequency as a sort of "natural language" messaging. They might broadcast or revoice the "voice" of hydrogen as a cosmically abundant element with a recognizable physical signature. To this day, it is unknown if what was called the "Wow" moment was in fact communication from aliens.

Meanwhile we also send out frequencies—radio and TV transmissions wander into space. Besides our media flotsam of sound, we broadcast intentional signals in the hope of being heard across vast distances of space and time. Giving voice into space says that we are here, these corporeal beings on this planet. Such human projection into the unknown is an intergenerational project, since even traveling at high-frequency radio speeds, a transmission today will take decades to reach any nearby stars. And should aliens hear it and respond, it may be another generation or more before we receive their message.

With expertise in psychology, anthropology, and animal biology, astrobiologists and SETI scientists speculate on how alien intelligence might communicate and how to best transmit to them. To break from anthropocentric thinking, they imagine what it would mean if alien bodies were more like that of a squid than a human. Such extraterrestrials might develop a base-eight number system rather than our base-ten system, which is conveniently connected to the number of fingers on two human hands. Then again, they might abandon anatomy and defer to the efficiency of a twelve-base system. Conjecture gets complicated when we realize that aliens may think and communicate with some sort of distributed cognition and distributed language that we could speculatively model from ant colonies or beehives.

"Tekeli-li! Tekeli-li!" is the haunting sound of alien voices in sci-fi horror author H. P. Lovecraft's masterwork *At the Mountains of Madness*. In this novella, Antarctic explorers at the farthest reaches of earth's extreme environment find a mountain-sized alien spacecraft that landed millions of years ago. While venturing into the labyrinthine city of the spaceship, the explorers encounter a variety of aliens and then try to escape with their lives from these shapeshifting, devouring creatures from the stars. Out of the depths of the spacecraft, the alien things cry in pursuit, "Tekeli-li! Tekeli-li!" Lovecraft borrowed the sound from Edgar Allan Poe's rambling tale of sea adventures *The Narrative of Arthur Gordon Pym of Nantucket*, where "Tekeli-li!" is the sound of a strange white bird and "Tekeli-li" is the sound made by the natives to the island when they see anything white brought on their island, which is itself devoid of whiteness except the birds who visit there. "Tekeli-li" becomes a figure for what cannot be understood and has no proper place within the culture. It is a vocalization marking alienness.

As with Kant's proposition, so often we see aliens as useful for us. But what if we were not the center and "Tekeli-li" is a marker for beings who live for themselves? In an episode of *The Twilight Zone* large-brained humanoid aliens wearing something like sparkly bathrobes have come to earth to share their advanced technologies with humans. On one occasion, after explaining their peaceful intentions, the aliens leave behind a large tome in their native tongue. The U.S. military deploys cryptographers. They make out the book's cover title: *To Serve Man*. Convinced of the extraterrestrials' benevolence, humans enjoy the new technologies, which remake the world into a planet without needs or wants. There is no more famine or war. Lucky groups are even taken to the alien ship, where, it is assumed, they are treated to pleasures that cannot be housed on this planet. Meanwhile, the 1950s TV audience awaits the twist they know will come with every episode of the program. As still more humans board the spacecraft, a cryptographer runs to try to stop them. He has deciphered the rest of the book and *To Serve Man* is a cookbook. The episode ends with the haunting lines: "Sooner or later, all of us will be on the menu . . . *All of us*." While we think we are connecting with them and understand their language, the aliens care for themselves and are indifferent to us as cattle for their consumption. This stands as a violently disjunctive element of our disjunctive synthesis with other beings. So much of them is different, alien. We are bit players, mere food, for the visitors to Earth who live out their own lives in which they are the most important beings. The universe is not here for us.

On the fortieth anniversary of the Voyager launch, William Shatner, Captain Kirk of *Star Trek*, was asked to commemorate the occasion with a message to be transmitted into space and directed at the Voyager crafts. He leaned into the NASA microphone and, with his signature performer's voice, proclaimed, "We offer friendship across the stars. You are not alone." We are, indeed, not alone. But will we offer friendship to the aliens among us?

The bodies open and project sound. But outside of understanding and beyond words, we can hear and become enthralled by the voice of terrestrial aliens. We play whale songs at night to go to sleep next to our purring cats or snoring dogs and awake to the chirps and twitters of birds. Their sounds reverberate in our ears, enter us, and draw us out to a vertiginous bridge between us.

TWELVE

Voice and Other Intelligences Continued

The Day of the Dolphin • A dolphin speaks • Floating minds and dropping acid • Interspecies connections • "Ways and means not previously conceived"

In the 1973 film *The Day of the Dolphin,* marine scientist Dr. Jake Terrell, played by the rugged George C. Scott, has made an incredible breakthrough: after years of training, he has taught dolphins English. With their heads poking out of the water and their toothy grins, they speak in high-pitched childlike voices that make them rather adorable but also creepy and sad, given the B movie–quality voice-overs. Thanks to their training, the dolphins even refer to the doctor and his wife as "father" and "mother," setting aside the biological meaning and treating it as a social description of the terrestrial-aquatic family. It is a paradise of humans and dolphins along the sandy coast and pristine waters of the Bahamas, but then the animals disappear. The dolphins are kidnapped, and an undercover government agent discovers that they have been taken by a shadowy entity called the Franklin Foundation, which, it turns out, has been using shell companies to support Terrell's research. The foundation plans to use the trained dolphins to deliver a package to a designated boat. Jake discovers the package is a bomb and the boat is carrying the president of the United States. Will Jake be able to stop the plot in time? I leave readers to discover for themselves how unwittingly revolutionary these dolphins may be.

The film draws its inspiration from the life and research of John Cunningham Lilly, who first investigated dolphin intelligence in the 1960s. It is thanks to Lilly that we have both the rainbow peace image of dolphins like on the TV show *Flipper* and the military special ops sort currently deployed across the globe.

Working in early neuroscience, Lilly performs what by today's standards seem like primitive and truly brutal tests, sticking electronic probes in the exposed brains of animals to see which muscles move and twitch. This is not impartial science. The military is interested in ways of infiltrating the mind of enemy agents and hoping to use brain implants to alter their behavior. After experimenting on a number of species, colleagues point Lilly in the direction of dolphins because they have the largest brain to body ratio of any animal except humans. But where to find dolphins ready for use? This leads Lilly to a Florida entertainment park, Marine Studios in Miami, where *Creature from the Black Lagoon* and other swamp-thing movies are filmed, and by chance where there are captive dolphins as part of the roadside attraction.

Amid the carnivalesque show, in an obscure, hot, and humid makeshift lab, Lilly begins his macabre experiments on dolphins. The day's selected subject of study is lured into a chute, restrained, and anesthetized, and has the skin around the skull peeled back and the skull cut open, probes at the ready. But time after time, the dolphins keep dying and the body count piles up. The park that sold him the dolphins gets increasingly upset. Finally, Lilly realizes that, unlike terrestrial mammals, dolphins cease to breathe when under full anesthesia. He begins again, this time using local anesthetics, but then something unexpected happens. He writes about the incident in his 1962 popular-audience book *Man and Dolphin*. As he begins the operation, the dolphin starts to vocalize, but not in dolphin sounds. According to Lilly, the dolphin mimics the voices of the humans about to operate on him. How can this animal know the need to project a voice across species? How is such a mind possible in a sea creature, and what is this aquatic mammal thinking? The scientist speculates that dolphins have a level of self-awareness and reflective consciousness equal to if not surpassing that of humans. He puts down the razor-edged scalpel and begins a new level of engagement that changes his life and the way we perceive dolphins.

How can we know what a dolphin is thinking or that he is thinking? A decade after Lilly's transformational encounter, the philosopher Thomas Nagel writes a much-acclaimed essay "What Is It Like to Be a Bat?" He explains that we will never know what it is like, since to do so we would have to be bats. But, if we've become bats, we can no longer communicate with humans what it is we're experiencing. Nagel's conclusion: not all facts are accessible to us. There are some facts that we know exist—that a bat feels something—but that we can never access. The dolphin being operated on bent his voice to mimic something he cannot know but conjectures is important to us—sounds from human mouths. Lilly becomes fascinated by a mind that can think outside of its own lifeworld to that of another species. In return, the scientist will spend the next decade trying to bend his mind to find a common consciousness with dolphins. In an attempt to twist free from Nagel's logical deduction that we cannot know what the mind of the dolphin is like, Lilly will resort to increasingly unconventional methods.

Moving to idyllic Saint Thomas in the Virgin Islands, the determined scientist sets up an elaborate seafront facility using U.S. Defense Department funding. The military imagines potential aquatic assassins; they foresee guards for ships at harbor and dream of scouts for discovering underwater explosive mines, and maybe even planting a few themselves. But Lilly is after something larger, grander. In *Star Trek*–like fashion, he wants to mind meld with dolphins to understand their way of thinking. This would crack Wittgenstein's lion problem, that if a lion could talk in human words we would not understand him because his way of thinking and deploying language would be foreign to us. Lilly formulates his line of questioning: "Having found a large brained species, we should attempt to determine whether its members have an intraspecies language. If we do not know that one exists, what anatomical and social conditions are most likely to make the existence of such a language highly probable?" Lilly, called the "cosmonaut of consciousness," suits up and begins exploring.

In a radical act of hospitality, or perhaps an exponential change in experimentation, or both, Lilly retrofits his 1960s mod home to accommodate aquatic companions. He has the concrete hammered out, molds made, and new concrete poured for canals and shallow pools running through the living room and study. Parts of the house are purposely flooded. Like oddly paired roommates, cohabitation may bring species closer together. Then he takes things a step further. Lilly sets up a pod-like container with body-temperature warm water and salt to facilitate flotation. He slips himself naked into this sensory deprivation tank and begins piping in the whines and clicks of dolphin vocalization.

The tank becomes another mind space. Floating amid dolphin pitches and screeches, Lilly hopes to experience an interspecies consciousness that can be transformed into human–dolphin communication. Still seeking the connection he intuited during his eureka moment with a dolphin in Florida, he begins to push himself and the animals further. Lilly calls upon the experimental investigative techniques he learned from the FBI and military psy-ops programs while working in his early animal cognition lab. One method stands out: mind alteration to reconfigure the ground of thinking and sensing. So, he acquires a few tabs of LSD—some for him and some for the dolphins. He enters the sensory deprivation tank and begins an interspecies acid trip. The imaginary soundtrack for this might be the 1960s psychedelic classic "Mind Flowers" by the East Coast pop-psych band Ultimate Spinach: "Sacrifice! Of your ego! Sacrifice, sacrifice, let it go! Let it go! Let it go!" Move toward the "myriad conscience of tomorrow's mind." Meanwhile, the dolphins were not consulted, and the drug likely hits them as waves of unwilled sensory distortions that they cannot stop riding for hours.

Based on Lilly's claims and ambitions in *Man and Dolphin*, anthropologist Gregory Bateson wrote "Problems in Cetacean and Other Mammalian Communication." (The essay can be found in his poetically titled book *Steps to an Ecology of Mind*.) Bateson mused that the human hand with opposable thumb has marked human evolution and consciousness but has also trapped us into viewing things in the world as objects to be manipulated. Since dolphins do not have hands, he reasons, they do not see all the world as objects for use. They are not stuck with our grasping mentality. Instead dolphins must have a more developed social mind, one that navigates the world by relating to others rather than manipulating them as objects. In a letter to Lilly, Bateson ponders, "If I am right, and they are mainly sophisticated about the intricacies of interpersonal relationships, then of course (after training analysis) they will be ideal psychotherapists for us." (Revolutionaries owe thanks to historian D. Graham Burnett for working through boxes of papers in archives to unearth this letter and for his published account of Lilly's project.)

In yet another immersive experiment, Lilly's lab assistant and researcher Margaret Howe Lovatt suggests that spending more time with the dolphins could offer a chance to bond with them. Lilly likes the idea, which reminds him of CIA intensive interrogation; so Lovatt and the young dolphin Peter live in isolation together six days a week for six months. Peter has been trained less than the other dolphins in the compound and so has less prior human influence. It is speculated that this might make him a better candidate for the experiment. So, the male dolphin and female researcher cohabitate. She sleeps suspended above the water at night and during the day does paperwork from a table just above the dolphin's aquatic world.

In this human-dolphin architecture and deprived of others of their own kind, Lilly thinks they may find a way to communicating and break the language barrier. Prior to the experiment, he had Lovatt read *Planet of the Apes* and prepared her to think outside of human cultural expectations. The goal is not to anthropomorphize the dolphin but to find an elusive common ground.

In working with Peter, the researcher discovered some unforeseen complications. Years later she explains: "Peter was a young guy. He was sexually coming of age and a bit naughty." Things got strange. By some accounts, Lilly had Lovatt wear red lipstick under the auspice of helping the dolphin see her lips when teaching English vocalizations. Lovatt recounts, "He was very, very interested in my anatomy. If I was sitting here and my legs were in the water, he would come up and look at the back of my knee for a long time. He wanted to know how that thing worked and I was so charmed by it. Peter liked to be with me. He would rub himself on my knee, or my foot, or my hand." Eventually the randy dolphin needed tending to if lessons were to proceed. After weeks of confinement, it appears that the dolphin's way of communicating was through sexual advances and, in an attempt to connect with the animal, the researcher lent a hand. "It would just become part of what was going on, like an itch—just get rid of it, scratch it and move on. And that's how it seemed to work out." A more lurid and far from accurate version would appear in *Hustler* in the late 1970s. As Lilly's array of unusual research methods became public, funding agencies abandoned him. Lovatt's experiment ended abruptly, but as D. Graham Burnett explains in "A Mind in the Water," "Lilly chalked it up as a victory for interspecies contact."

In his preface to *Man and Dolphin*, Lilly is wide-eyed about the latent possibilities of dolphin research. He sees it as a stepping-stone to cosmic ambition: "In a way this is a crude, elementary handbook for those humans who are interested in the realization of such communication. If no one among us pursues the matter before interspecies communication is forced upon Homo sapiens by alien species, this book will have failed in its purpose. But if this account sparks public and private interest in time for us to make preparation before we encounter such beings, I shall feel my time was well spent in the research here described." Lilly's research methods were strange and unethical. Their outlandishness only further emphasizes the complexity of bending the human mind to understanding another intelligence. As he says in *Man and Dolphin,* he was driven by an encounter and the promise that nonhuman life brings new worlds to our shores: "We shall encounter ideas, philosophies, ways, and means not previously conceived by the minds of men."

THIRTEEN
The Crack

Revelations in the quiet of the night • Snuggly confinement • "I'm a bee" • Guadeloupe raccoons • Rediscovered reverence for the bald eagle and Chinook salmon • Forgotten dreams • To come

"Goodnight dear readers. Go home, lock your cage well! One never knows what may happen. Sleep with one eye open. At any rate, sleep well. Pleasant dreams." And so ends part one of J. J. Grandville's satire of Victorian manners in *Public and Private Life of Animals*. Let us consider what might happen in the lull toward sleep.

After a day of manufactured wonder at Disney World, a child puts her head to the pillow, thinking of all the magical events as if they are still happening in the commercial utopia. Visiting a national park, this same child might draw toward sleep imagining the roaring waterfall or some timid and unusual wildlife. At a later age, about the time one begins to disbelieve in Santa, she may go to sleep realizing the Disney machines shut down and performers go home, but the national park waterfall does not turn off, and hasn't since well before she was born and won't perhaps for long into the future. And the wildlife? They or their offspring are still there. They are not performing and are not paid. The animals live beyond us.

The girl may think—as children do—that adults do not know this. It is a secret she alone has discovered. Like a door to Narnia, it opens to a place that older humans do not understand because they are preoccupied with jobs and houses and groceries and being adults. This secret is not simply that animals have worlds beyond our grasp. Rather, these worlds reflect back to us an image of ourselves that is unrecognizable. We are not the civilization we think we are, and we are not who we think ourselves to be. With wonder and a measure of cautious apprehension, the child stepping through the door to animals' worlds returns with a crack in her perception of humans.

Adults sitting leisurely on a sofa with feet propped up and with especial cozy socks designed for evening reclining can feel rather comfortable and secure. Bricks and drywall provide insulation from the unpredictability of the great outdoors. Later in the evening, the seat becomes uncomfortable and the socks too warm. The walls close in, as does a general feeling of unease. On yet another night, the eerie feeling returns but with more focus this time. A claustrophobia. We have created cages for ourselves. Our range of motion becomes limited by the confines of self-domestication. All the while, the child with a crack in her perception feels nature going on without us. Slipping from the domestic zoo, her mind goes feral.

Maybe it starts with the stuffed animals that swarm her bed, or it begins with the phrase "Reindeers are better than people" that still echoes from repeated watchings of *Frozen*. The crack appears and then untethers her from commercial interpolation of the animal kingdom. Creatures lure the girl. In such moments, her legs alight across the yard, earnestly moving at her very highest speeds while repeating "I'm a bee. I'm a bee. I'm a bee." The intensity of the child's insistent repetition and the effect of her gait make the divide between humans and animals more of a mobile border. Then, the border that was once a divide becomes an affinity that she can leap across. Sometime later, returning to her little-girl self, she feels not herself as a daughter or student. The ways bees see her haunts and slightly unhinges her place in the civilized world.

With a furry gray and black body, a ringed tail, and its signature black mask, the raccoons of Guadeloupe are a lure for the residents of the Caribbean Island. They roam freely, stealthily pillage farm fields, root through garbage cans, and prey on smaller animals. In other parts of the globe, the omnivorous raccoon is dubbed a pest. And certainly their bandit masks do little to help their image. But here, despite the raccoon's rambunctious behavior, the people of Guadeloupe have developed a special relationship to the critters.

The advent of this relationship can be traced to the Smithsonian biologist Gerrit Smith Miller Jr. *Radiolab* journalists report that, in 1911, a small box from Guadeloupe arrived on Miller's desk. Inside was a tattered, dead raccoon. He dutifully examined the body and recorded measurements, weight, and notable characteristics. At eighteen inches, this specimen was

smaller than the standard two-foot, eighteen-pound version that roams North America. Miller tagged and logged the new arrival as a different species of raccoon and called him *Procyon minor.* Because the creature was only found on the islands, authorities there gave the raccoon special standing and made killing, capturing, or owning a "Guadeloupe raccoon" illegal. Islanders took great pride in their *Procyon minor.* Crops ruined, henhouses raided, trash strewn across streets, and yet the Guadeloupe raccoon remained revered.

Ninety years later, renowned mammologist Kristofer Helgen was doing comparative work on members of the raccoon family. He accessed the Smithsonian library of preserved animals and unboxed Miller's tattered raccoon hide. It was indeed smaller than the North American variety. But then he picked up the skull at the bottom of the box. He examined the teeth and jaw and then the frontal bone. There he noticed spacing in the sutures of the skull. The bones had not fully fused, meaning that this was a juvenile animal. Its small size was not due to being a unique species; simply, the animal was not fully grown. Helgen is responsible for removing the Guadeloupe raccoon from its esteemed position as *Procyon minor.* When news of the finding reached the island, the change in scientific status made no difference. People there still believe their raccoons are special and the laws against harming them remain in place.

There are moments in which civilization is not immune from the animal lure and where boundaries between us and them become affinities thrown across borders. Citizens of Guadeloupe have connected with this animal in a way that defies scientific reasoning. Instead, affections allow the islanders to slip into another way of seeing the world. The island belongs to the raccoons as much as to the human inhabitants.

Think of the bald eagle and its varied history in the United States. While Americans see it as a graceful and powerful predator, it can also be depicted as an opportunistic feeder and scavenger. Fifty years ago, bald eagles were near extinction because of habitat destruction, the use of DDT pesticide, and illegal shooting. Then flags were waved and concerns for a national symbol were invoked. Americans were asked to amend their habits. Some people changed reluctantly and resentfully and only because of new federal laws, while others did so with remorse for human ecological harm. The changes in human behavior made a difference. Fast forward several decades, and today Americans say with pride that their national bird, the spirit of the nation, now thrives.

At the San Francisco airport baggage counter, twenty-four members of the Winnemem Wintu tribe check a drum, manzanita logs, a container of sacred water, and ceremonial spears, bows, and arrows. Then they board for an eleven-thousand-mile flight to New Zealand. It is a spiritual mission to ask the salmon to come home. The Shasta Dam blocks the fish's access to the tribe's land along the McCloud River. Chinook populations are dwindling, as is the cultural way of life of the tribe. Decades ago, when salmon were plentiful, some fish were flown to New Zealand so that the island could develop its own fisheries. Now Winnemem Wintu tribal representatives meet with local New Zealand Maori leaders. They perform ceremonies asking the salmon's forgiveness for not doing enough to stop the dam construction. They show remorse and ask the salmon to come home. And then, putting action alongside their prayers, Winnemem Wintu leaders secure salmon eggs with the hopes of reconnecting with the fish and reviving tribal customs.

While dams or DDT or species misidentification are modern problems, beneath them is an almost timeless tension between warding off animals and being drawn to their lure. Resting your head upon your pillow and blanketed in comfortable sheets with calculated thread count, you may feel civilization slip from consciousness and then the haunting and then the tug and then the fall to a more-than-human world and forgotten dreams.

In the mineral-stained limestone cliffs above where the Ardèche River once flowed, there is an almost imperceptible crack. Stepping across its threshold, one walks from the sunlight into the darkness of a narrow chamber that tilts precipitously downward. Descending, the cavern opens in all directions. Stalactites and stalagmites populate remote corners and calcite walls sparkle. Farther along, in a large chamber, images and flattened figures appear on the walls. A single bear is outlined in ocher and then another. One or two bison reveal themselves in charcoal black and then herds show up as if they are roaming the open plain. Layers of horses stacked one upon another seem to gallop with their motion caught by repetition of line and pattern as they push from folds in the stone. The great curved horns of an ibex move along the side of a rock, then its full, corpulent body comes into view. Simple, strong lines accurately render lives that have unfolded on the other side of the darkness.

Out there, animals roam along forest and valley. Humans carefully watch and carry their sightings into the cave, and then the animals are there, transposed onto stone walls. The outside is folded inward—from observations in the field to the caves of the mind, then in the darkness of the stone cave, and into the culture of a people. Thirty-five thousand years later, the animals are here in the minds of contemporary visitors.

While access to the Chauvet Cave in France remains heavily restricted, filmmaker Werner Herzog was permitted to make a movie that would give the rest of the world a glimpse of this other space and time. With minimal crew and restrictive filming conditions, Herzog created *Cave of Forgotten Dreams.* Today, we can see the primordial cave and watch the animal paintings in darkened theaters where moving pictures are projected on the walls.

The film nears its conclusion as the camera lingers on images found in the farthest chambers of the cave. Past a bear skull placed tens of thousands of years ago as a totem or sacrifice on an altar-like stone slab, and then into the far recesses of the End Chamber, a well-wrought bison looks into the distance. Following the animal's gaze to the Sacristy, there hangs a column from the ceiling with a curious image on it. On the back side is a bison's head in profile and featuring a singularly attentive white eye. The line from its neck extends into a human arm, hand, and fingers. Below the bison-human is a woman squatting with legs spread open and straddling the stalactite like a phallus. Animals move outward from the woman's loins as if she is endlessly giving birth to them.

If asked when the revolution will come, animals would likely reply that this human question aims too low. Like the infinitive "to come," the revolution is ongoing and almost suspended in time. The animal revolution has come, is coming, and will come. It is here in a long temporality beyond the arc of human history.

FOURTEEN
Bearing Witness

Hair • Sasquatch • Resonant relations in the open • The intoxication of goats • Dwelling differently • "I believe"

> The division of life into vegetal and relational, organic and animal, animal and human, passes first of all as a mobile border within living humans.
>
> —Giorgio Agamben, *The Open*

In the film *Memento*, because of a recent brain trauma, the protagonist cannot retain events in his long-term memory. This makes it difficult for him to track down the unknown person who killed his wife. He wakes up forgetting if he has a wife and forgetting his intention to find the killer. His solution is to tattoo her death and the facts of the case on his body. His skin becomes the outward sign of the things his inner mind cannot contain. The clues are inked into him. While the movie had a lot of plot problems, I always liked the almost religious figuration of the tattoo as an outward sign of an inward symbol. But I never got a tattoo. I'm covered in hair—forearms, biceps and triceps, shoulders and back, legs and most definitely chest. Language and tattoos as tools for memory and meaning never cohered with the animality of my body. Hair as the friction to reading and writing prevented my self-inscription. Once on a beach, I took my T-shirt off and one of my very hairless female friends gasped. She was not amused. I went to the Natural History Museum in New York with their infamous dioramas.

Looking at the mock-up life-size model of a Neanderthal, I whispered, "He's got nothing on me." My outward sign is the hair of animality. I awake every morning and dress aware of my body in the world, one that has to be covered and fitted. Hair tucked into my shirt and the extra-high shirt button to keep it down. Then I can be like the other people at work, the ones whose bodies do not give offense by their animality.

More recently, I have adopted the Sasquatch as a comrade. I don't know any Sasquatch yet, but I talk to them when alone: "You understand, you outcast and hunted malformed being somewhere between beasts and humans." I am convinced that the sustained search for "Bigfoot" is so that one day, when captured, the hominoid can profess to people what it is like to be "the missing link," to have a nonhuman animal comportment, an intelligence that can communicate this animal nature to us. Or maybe Sasquatch is the external sign for us all, a collective cultural sign for our own animality and our desire to commune with it. The aura of our bestial animal nature is there, illusive and blurry on the horizon.

We need Sasquatch. This being serves as a reminder and marker of the resonant relations between our animality and the animality of the nonhuman. We forget our animality that remains forlorn, choked off, stifled. It grows stunted and deformed, but it does grow in us and reveals itself at untimely moments. If we are to decrease the abyss between ourselves and other animals, and even the mobile border of animality in us, we need ways in. We need ways of finding affinity and means of remembering encounters.

An encounter with a nonhuman animal opens something in us. Sure, there are animals domestic and wild all around us twittering, barking, purring, snuffling, and padding along. We see them and recognize them but rarely do we take the opportunity to encounter them. We are busy and an encounter means a pause or even a tear in the fabric of the mundane. Who can make time for that?

Sometimes the incidents of revolution do not ask us. They

come uncalled for and unabated to break through our complacency. But even without such insistent moments, the animals are out there in the long trajectory of the revolution. We can provide an opening for these events, these encounters with animality, these moments of resonant relation when we are pulled from ourselves into a third space, a disjunctive conjunction where, disjoined from our world, we find there is a larger open field of play amid us and them. And in this pasture or parking lot or backyard or remote woodlands, we find that our own corporeal being with its frictions and physicality is in a conversation with fur or scale or feather or claw or hoof or beak or tusk or dark eyes. A metonymic part of the animal becomes a lure, and then we find ourselves in a still pause and drawn into another animal's relationship to the earth. Standing there in our animality with the grit of the world at our feet, we share that dirt of the earth and feel a thickness to the air between humans and animals. The human world will return and the animals' worlds will recede, but such events are an opening of awareness that lingers.

Once, in a moment of mourning, I shaved my hair. All of it. As if something deep was lost to me and new growth would come from shedding it. I saved some of the hair and put it in a handcrafted wood bowl with its grain worn smooth from years and use. Then I headed to the mountains. At the height of fourteen thousand feet, well above the tree line along the Continental Divide, geological time is palpable in every boot-to-rock step and every expansive view. Along the shoulder of the mountain, walking the path on my descent was a herd of mountain goats. It is a rare sighting in these mountains. Their dark, unreadable eyes watched me. They walked on and I cautiously walked behind them, giving them ample distance. Along the way, I collected their shedded strands and tufted balls of hair. Once home, I put these weathered mountain tufts in the smooth wooden bowl. There, my black hair rested against the white goat hair. Threads mingled. It was not a popular item for visitors to my house, but it made sense to me and gave me peace. These were artifacts from moments. They are markers and reminders of affinities between bestial beings. Why couldn't they see that? Here, fragments from a distant mountain beast meet the distant fragments of myself. The threads talk by touching. Woolly, clear-aired space lent something to my earthbound urban life, my social being, my awkwardly weighed inoperative communing with other humans.

Sometime later, a dear friend gave me a thrift store coffee mug with a poorly rendered goat embossed on it. The goat was not quite right. It was not a satyr but as if an artist who only knew how to draw humans was tasked with rendering a goat and as if they had never seen a goat before but had to form it from secondhand descriptions. Despite the legend of Ethiopian goats discovering coffee berries that would somehow redeem this ill-shaped goat mug and make it an icon of caffeinated consciousness, I only drink whiskey from it. And with the whiskey comes the intoxication of bodies traversing mountain peaks with no regard for the world below. While mistaken for a romantic construction, it is my way of knowing without language. I don't want a rational discourse that drags animals into the human world as trophies of human dominion or icons of cuteness. I long for an expression and marker of remembering worthy of the encounter.

Encounters change the state of affairs. I've seen things . . . and that changes everything. Such moments do not have to be fleeting but can be a part of the stories we tell ourselves about ourselves. They can be part of who we are. This changes the narrative for culture. As philosopher Martin Heidegger says: before we build, we dwell. In other words, every building is constructed based on our current ways of dwelling, and our ways of dwelling tell us something about what we think and value. Taking encounters with animals seriously and then expressing these moments without co-opting them, we can change the way we think and dwell. And that will change the way we build, not only homes and offices and megastores but how we build policy and laws.

In building we translate—even awkwardly—the space of encounter so that we might structure the human world to accommodate third spaces—places open and hospitable to other animals seen and unseen by us. To make such space viable and to remind ourselves of encounter, we can perform small ecological rituals of daily life that do something for the shared earth. We can fashion mementos to help us remember. To recall this open space, we need occasional reports in the news that encounters are still out there and still possible and ongoing. We circulate YouTube videos of a pigeon who runs off with a video camera and films from the air or an octopus who tries the same stunt underwater resulting in avant-garde, nearly Jean Painlevé film. The videos aren't the encounter but are a marker for memories. We need these reminders that renew our efforts to dwell differently and prevent human usurpation of the shared earth.

"I believe," say those who look to the stars for alien life or peer deep into the woods for signs of Bigfoot. We have an array of cultural myths about animals. Believe these myths and take them seriously. Activate them. They are not allegory or fantasy. They are here and now. They are values and ethics and actions. Science has its role; we fashion wonder from images of giant squid brought to us from research vessels that plumb the depths of the oceans where light never penetrates. We marvel at extremophiles that live in boiling waters or in frigid temperatures nearly a mile below the Antarctic. Magic has its place as well. Riding a horse is an enchanted act such that one can begin to believe in centaurs. Standing at a precipice of a high desert plateau and watching a California condor dive from heights, extend its nine-foot wingspan, and swoop inches from one's head, one can feel for a moment the wonder of flight, not abstractly, not at a distance, but as if transmitted from bird to human.

The animal revolution hacks into what it means to be human and opens us to recognize animality within us and out there. They beckon. Believing and heeding their call can change who we are, can change the human world, and transform the earth on which all beings dwell.

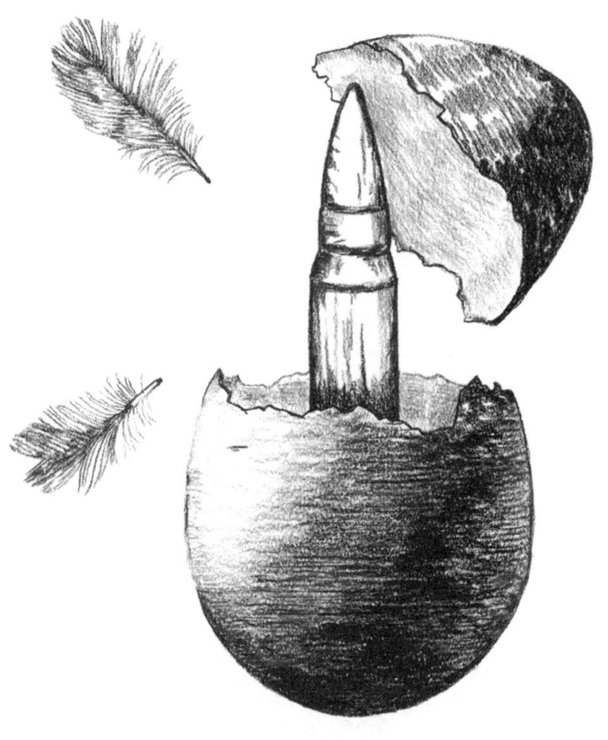

CODA

Why We Need Better Stories

Condor and lead • Breeding in captivity • Ted Nugent and a metal less heavy • Cold-water fish • What humans value • The stories we tell

A California condor spreads its massive wings and catches an updraft along the Grand Canyon. With keen eye it searches for carrion in the canyon's folded, descending slopes as its ancestors have done for tens of thousands of years. More than a mile below, a small brown fish with a large hump between its head and dorsal fin swims idly in the rivulets of a tributary of the Colorado River. The humpback chub have swum in these waters for thousands of years. With clear blue skies of the Southwest, massive geological formations, and deep running waters of the Colorado, it would seem that in this remote desert environment animals carry on with little notice of bustling human civilizations. If only it were so; the peaceful harmony of nature is mere camouflage. The California condor and humpback chub are endangered species because of human modifications of their environments.

In 1987 California condors were in dire shape. Their total population fell to a mere twenty-seven birds. Their leading cause of death was lead poisoning. Much like eagles and vultures, condors scavenge for food with a particular preference for the corpses of large mammals. When hunters gut and field dress beasts they have killed, shards from their lead bullets sometimes lodge in the innards left behind. Scavenging birds consume the flesh but also, unfortunately, the lead. For several human generations we have seen the effects of lead introduced into ecosystems by fishers and hunters. This is why anglers have shifted to nonlead sinkers and duck hunters use nonlead bullets.

With the condor, their numbers were so dramatically low that the last of their kind were apprehended and brought into a captive breeding program. Normally a condor will fly a hundred miles or more each day in search of food. But now the last of their kind were confined to enclosures at the San Diego Wild Animal Park, the Los Angeles Zoo, and affiliate human enclosures for animals. The ancient beasts folded their wings and awaited their fate.

Eventually enough birds reproduced that some could be reintroduced into the wider world. Today, with just over five hundred birds in total—a third in captivity and the rest in the wilderness—the program is considered a model of success. But even the "wild" birds are far from any common idea of wild. Their wings carry large numbered tags so that they can be found by radio location devices and identified at a distance. Annually, each bird is captured, weighed, and given a blood test for lead levels. If a condor is found to have dangerously high amounts of lead, they are taken to remote facilities—makeshift condor hospitals—for blood transfusions and then, after weeks of monitoring, released. Caring for condors is a full-time job for field biologists living and working in remote parts of California, Utah, and Arizona.

To further help the condor, there has been considerable effort by Game and Fish departments to eliminate lead bullets, especially in the condor's domain. But hunting laws vary by states, and the condor roam across state lines unaware of the differences. In California it is illegal to use lead ammunition in designated sensitive wilderness areas. In Arizona and Utah, use of nonlead bullets is encouraged but not mandated by law. Policy makers in these more libertarian-leaning states observe that telling hunters what to do can cause a backlash, so it is more effective to engage their sense of responsibility. Heavy metal rock star Ted Nugent sells a line of nonlead patriotic bullets complete with flag imagery on every box, or as one reviewer writes: "Great ammo from a true patriot!" It is still too early to tell which avenue works best, regulation or appeals to a hunter's stewardship ethics.

Meanwhile, in remote Bright Angel Creek tributary of the Colorado River, Grand Canyon National Park field biologists remove trout from the water to create a sanctuary for the chub. They are trying to solve a problem that started half a century ago. In 1966 the seven-hundred-foot-high concrete wall that is the Glen Canyon Dam was completed, and with it came large-scale ecological devastation including effects on the indigenous humpback chub. The fish had evolutionarily adapted to the warm, muddy waters of the Colorado River. When the concrete interrupted the Colorado, it also changed the quality of the river by releasing cold and clear water from the bottom of the dam. The chub could not adapt to the change and reproduction levels dropped dramatically. Apparently for the chub, sex in cold water was not all that appealing. Adding insult to injury, the Park rangers added trout to the river because, unlike the chub, trout are well suited to cold, clear waters. But trout eat baby chub. The result was a dramatic decline in the indigenous humpback. Off-site fisheries were enlisted to keep the population viable until a solution could be found.

With boats, waders, and buckets, environmental biologists started netting fish, identifying the chub, and transporting them to rivers and streams more akin to their original home and more or less free of trout. The Little Colorado River now has a few thousand chub. And in 2018 after more waders and buckets and nets, this time to remove trout from Bright Angel Creek, chub were set loose in their native Grand Canyon. As artists Bryndis Snaebjörnsdóttir and Mark Wilson show in their exhibition *Trout Fishing in America and Other Stories*, human reaction to the return of the chub is revealing. Conservationists and field biologists were happy to see the fish's restoration to its pre-1960s habitat. However, anglers who want the pleasure of catching game fish deep in the interior of the national park were strongly opposed to change. After all, they argued, why are scientists imposing their will on the trout already here benefitting the park as a recreational, taxpayer-funded, public space? For a more general public, the value of the project seemed meaningless. Why spend federal dollars on a fish when so many children need food security and a better education? Little known to the public, the funds were skimmed from profits generated by the dam's hydroelectric energy. Knowing the dam would cause inevitable environmental destruction, ecologists argued for a built-in mechanism to fund some minimal remediation of the dam's ills.

These stories are messy. Condor are endangered by humans, then we invest in work-arounds to bring back some minimal survival of the species. Use of nonlead bullets depend on legalization, goodwill, and patriotism. Humans built the Glen Canyon Dam for the selfish ends of accessing more water and power for cities in the Southwest. But to mitigate these environmental sins the power company pays a tithe for ecosystem restoration. Park rangers are reintroducing the humpback chub, but it was rangers in the 1960s who introduced the trout

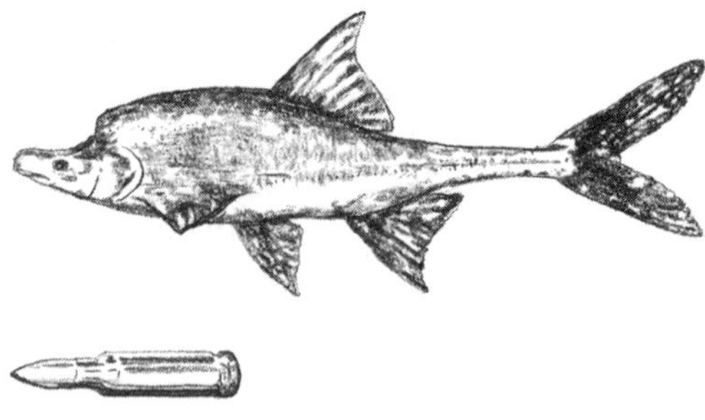

that helped endanger the chub in the first place. Those who see the national park as a monument to preservation support the return of the chub while those who see the park as a site for recreation advocate for trout fishing. And for most citizens the whole issue of why we invest in animals is up for grabs. The story of the condor and the chub is repeated with different animals and ecosystems in their own particular variations across the globe. Interventions and counterinterventions. Stopgap measures, delays, revaluations. Scales of domination and scales of restitution.

The animal revolution is a persistent and sometimes violent question: can we care beyond ourselves? The cases of the condor and chub provide a mixed review of human answers. Many career scientists, park rangers, and volunteers dedicate themselves to the flourishing of nonhumans. Then there are people who want dominion without burdens of responsibility—such as the hunter using lead bullets in ecologically sensitive areas and the angler who could fish for trout in many places but also insists trout thrive in the national park. Sometimes voluntary measures and appeal to stewardship work and in other cases we need legislation to correct for our tendencies toward selfishness.

As cultural theorist Donna Haraway explains in her aptly named book *Staying with the Trouble*, "It matters what stories we tell to tell other stories with.... It matters what stories make worlds, what worlds make stories." To be an ally of the revolution means telling stories of animals as active and important players on the earth and in our own social worlds. The animals keep reminding us of that. They rupture the stories we tell ourselves and give us new stories. More profoundly, the revolution changes the frame; it is not just where animals fit within our stories but where do we fit within theirs? Are we staunch allies or wayward, occasional advocates? Or will we be close-minded seekers of our own safety—animals be damned—or, beyond safety, will we be seekers of dominion on this earth? Or perhaps we are each and all of these at different times and in different places. If we are going to change things voluntarily or through legislation and if we will change consciously or at deep unthought habitual and structural levels, then it begins by transforming the way we think about ourselves. We need better stories about who we are—the we of we and who is included and excluded. There are untold incidents of animals in revolt every moment of every day. Find them, tell them, weave them into how you see yourself, how you see the world, and how the world sees you.

Solidarity, comrades.

AFTERWORD
Beasts of Sorrow

Eugene Thacker

A cursory glance at the representations of animals in literature, art, philosophy, and science can give one the impression that what makes human beings unique is the deep anxiety with which we obsessively represent animals to ourselves. From medieval bestiaries to DNA databases, it seems that we tend to view animals in one of two ways: literally (natural history, biology, genetics) or metaphorically (literature, philosophy, the arts). Linnaeus or La Fontaine. Food or pets. Even within the cultural sphere, animals are ceaselessly made to perform specific kinds of works for us as human beings: human beings are satirically analogized as animals; animals, we are told, mirror who we are as human beings; philosophers ruminate on the animal as some primordial ontic base from which "the human" ascends; and in our rapacious capacity for techne, animals have become unwitting participants in the human-centric shaping of the world. It is as if to suggest that the human being is simply the animal that *worries* about "animals"—and whether, after all is said and done, the human too is "merely" an animal.

Ron Broglio's *Animal Revolution* holds human beings accountable for this myopic, dichotomous approach to animals. There is always something else, or something other, when it comes to the animal. There is no "animal" without the human to name it as such; animals themselves could not care less. It is, perhaps, this *opacity of the animal*—precisely because we both are and are not animals—that defines us as human beings. Nietzsche's horse, Sōseki's cat, Lispector's roach . . .

In 1859, Arthur Schopenhauer published his final book, an unwieldy six-hundred-page, two-volume work of philosophical odds and ends. Actually, to call them "philosophical" is being generous; the writing is mostly aphorisms, fragments, observations, anecdotes, rounded out by healthy doses of vitriol for philosophers, professors, priests, politicians, writers, readers, the public, people who make loud noises, and, well, just about everyone. By now, the curmudgeonly misanthrope of Frankfurt had all but given up on receiving any recognition for his work, and this last book seemed at once a final defiance and yet also a secret plea. Such a simple little thing, to be recognized by one's own kind. As Schopenhauer himself once noted, the misanthrope actually needs others; misanthropy requires company.

Schopenhauer capped off this massive, doorstop of a book with a short parable:

> One cold winter's day, a number of porcupines huddled together quite closely in order through their mutual warmth to prevent themselves from being frozen. But they soon felt the effect of their quills on one another, which made them again move apart. Now when the need for warmth once more brought them together, the drawback of the quills was repeated so that they were tossed between two evils, until they had discovered the proper distance from which they could best tolerate one another. Thus the need for society which springs from the emptiness and monotony of the lives of human beings, drives them together; but their many unpleasant and repulsive qualities and insufferable drawbacks once more drive them apart. The mean distance which they finally discover, and which enables them to endure being together, is politeness and good manners. Whoever does not keep to this, is told in England to "keep his distance." By virtue thereof, it is true that the need for mutual warmth will

> be only imperfectly satisfied, but, on the other hand, the prick of the quills will not be felt. Yet whoever has a great deal of internal warmth of their own will prefer to keep away from society in order to avoid giving or receiving trouble and annoyance.[1]

Frequently referred to as "Schopenhauer's porcupine," this little parable addended at the end has since become the most popular passage in the whole work. With the porcupine parable, Schopenhauer immediately suggests a relation between misanthropy and animality. On one level, the parable uses animals as analogies for human behavior. What the porcupines do "naturally" (that is, unconsciously), we humans do quite intentionally (especially when proper manners are at stake).

But, rereading the parable, I can't help but notice how Schopenhauer makes it seem that even the porcupines get irritated with each other, as if they are perpetually caught in a tedious tragicomedy of being alone together or together alone—oh, you again, nice weather we're having, isn't it? Is this irrevocable back and forth a form of suffering specific to human beings? Or is it something that is common to all animals, inclusive of human beings? The tired drama seems to go on interminably until we die, in which case the terms *alone* and *together* cease to have meaning.

What's more, Schopenhauer seems to offer a bit of (unsolicited) advice at the end: be neither a human nor a porcupine; leave the drama altogether. This is unsurprising, given Schopenhauer's affinity for the ascetic tradition of hermits, mystics, quietists, and the like. But he also knew this was impossible. Even Antony, in his desert cave, was beset by demons (the noisy neighbors of fourth-century Egypt, one would suppose). And if it wasn't the demons, it was the long line of acolytes making pilgrimage to Antony's cave, hoping to receive helpful tips on how to be a hermit. In a way, the real takeaway of Schopenhauer's parable is the insufficiency of becoming-animal *or* becoming-human.

That said, Schopenhauer also seemed to have a special affinity for dogs. He famously had a series of small pet dogs throughout his long life and named them each Atma. An avid reader of the Upanishads, one would suppose that when one dog died, it was simply reincarnated. I have had many dogs, and yet I have had one dog. Although Schopenhauer railed against humanity throughout his writings, he was also known for his indictment of animal cruelty. In one passage he writes:

> And the highly intelligent dog, our truest and most faithful friend, is put on a chain by us! Never do I see such a dog without feelings of the deepest sympathy for it and of profound indignation against its master. I think with satisfaction of a case, reported some years ago in *The Times*, where Lord —— kept a large dog on a chain. One day as he was walking through the yard, he took it into his head to go and pat the dog, whereupon the animal tore his arm open from top to bottom, and quite right too! What the dog meant by this was: "You are not my master, but my devil who makes a hell of my brief existence!" May this happen to all who chain up dogs.[2]

The passage occurs in an essay on the topic of human suffering. Schopenhauer not only expresses his disgust toward animal cruelty but in a way has written another parable, one about the human condition. Years later, the Japanese poet Sakutarō Hagiwara, himself an avid reader of Schopenhauer, would write:

> I do not know where I'm going,
> a large, organism-like moon is vaguely afloat ahead of me,
> and in the lonely street behind me,
> the tip of a dog's thin long tail is dragging on the ground.[3]

As if to keep dogs to their word (. . .), it is said that when Schopenhauer was out walking his dog Atma and it was misbehaving, he would scold it: "You are not a dog, but a human! A *human!*"

Notes

1. Arthur Schopenhauer, *Parerga and Paralipomena*, vol. 2, trans. E. F. J. Payne Sato (New York: Clarendon Press, 2000), §396, 651–52.
2. Schopenhauer, §153, 297.
3. Sakutarō Hagiwara, "Unknown Dog," in *Cat Town*, trans. Hiroaki Sato (New York: NYRB Poets, 2014), 44.

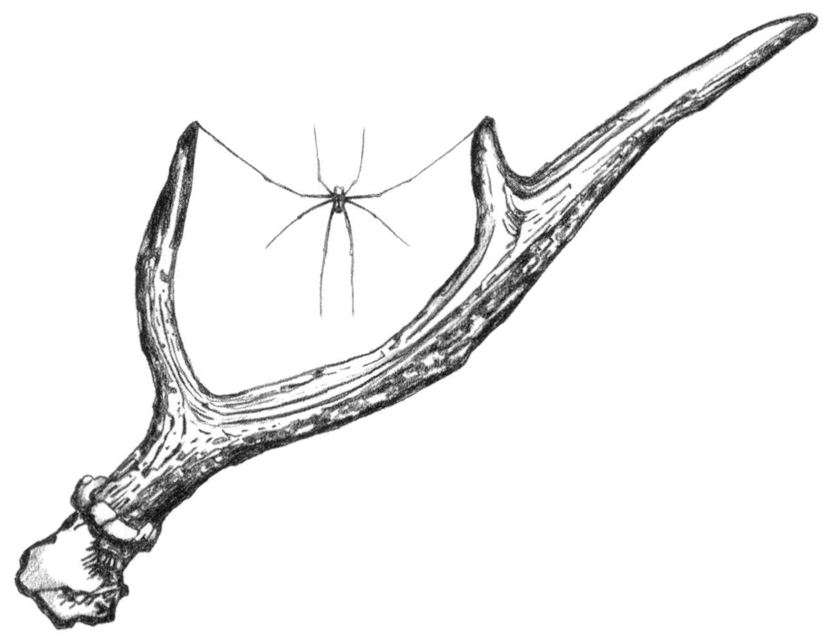

ACKNOWLEDGMENTS

I began thinking of the revolution eighteen years ago inspired by the work of comrade Fredrich Young and by early animal studies scholars such as Susan McHugh and Steve Baker. Since then I have found many allies while traveling the long and winding path toward this book. Artists Bryndis Snaebjörnsdóttir and Mark Wilson have been important companions in making and thinking with me from the deserts of the American Southwest to remote corners of England and Iceland. I'm grateful to Cary Wolfe, who saw the mountain goats and aided me and my ideas while hiking the dizzying heights of the Colorado Rockies. I first brought the revolution to Doug Armato as a magically unreal proposition many years ago. I am thankful for his faith in the strangeness of the work and his insistence that now is the time to get the writing done. I am grateful to Marina Zurkow for keeping wonder alive in her art and bringing such energy to the images that are in dialogue with my writing here in *Animal Revolution*.

I would not be able to think or write without those friends and allies who live with passion the worlds they write, make, and bring to life: Margret Grebowicz and her spellbinding experimental phenomenologies; Tracy McDonald and her punk take-on-the-world comportment, David Clark with such insight and care; Giovanni Aloi and his provocations and his fabulous *Antennae*; Kira O'Reilly, who thinks with animality and bodies; Eva Hayward, who writes magic and lives it too; Eugene Thacker and his faith in my work and cosmic pessimism; Stacy Alaimo, with kindness and bravery in living and writing; Kari Weil, whose writing resonates year after year; Erika Hanson and her continued curiosity; Krista Davis queering worlds; Val Lyons, who channels Joseph Beuys; Nadine Boljkovac and her time-image; Anthony Pessler and his animal icons; Claudia and Michael, who live intensely; Sally Ball, who makes poetry; Maria Whiteman, who persists in creating; Lynn Turner and Undine Sellbach, who seed the revolution with their work; Joanna Grabski, who practices companionship with Leo; Donna Haraway, who is a true field marshal of the revolution; Kate Ready and her quiet, persistent being; Devoney Looser, with her thoughtfulness at all times and her felt scholarly passion; Celina Osuna, who lives music and deserts; Adam Nocek and Stacey Moran and their academic and personal hospitality; Frida Beckman and her brightness of thought and being; the whole gang of the SLSA (too many to name), who make a home for novelty and oddity of ideas. And thank you to those out there in the world who go unnamed here but who have animated the way I live and the writing of this book.

I continue to be grateful to Arizona State University for providing me generous space for thought and writing. I appreciate the leadership team of the English department for all they do in facilitating research. And I am deeply grateful to the staff of the Institute for Humanities Research, who consistently give their skills and energy to making work happen; thank you Liz, Lauren, Barbara, Sarah, and Celina and to Elizabeth Langland, who mentored me at the Institute.

I could not write nor could I live in this world without my family, some of whom are still with me and others who have helped me along the way. So, thank you Ty and Alma. Britt and Marina, your joy has seen me through. Ayanna and Jonathan, you have kept me going. Jane and David, you are there even at a distance. Thank you Forrest, who carries worlds and gives me reasons to write. Alexander, you have kept me moving forward. Barley, Barlz, you have been a true companion. Marcel, I am thankful for you pulling me out of the abyss, repeatedly. Jeffrey, thank you for continuing to believe in me and reminding me what friendship can mean. Matt, you taught me how to write this book, and you share the intoxication needed to have these visions.

FURTHER READINGS

Simon Adler, "Stranger in Paradise," *Radiolab,* January 27, 2017, https://www.wnycstudios.org/podcasts/radiolab/articles/stanger-paradise.

Giorgio Agamben, *The Open: Man and Animal.*

Giovanni Aloi and Susan McHugh, in *Posthumanism in Art and Science.*

Fahim Amir, *Being and Swine: The End of Nature (As We Knew It).*

Steve Baker, *Postmodern Animal.*

Matt Bell, *Appleseed.*

Joseph Beuys, *I Like America and America Likes Me.*

Rodney Brooks, "What It Is Like to Be a Robot," https://rodneybrooks.com/what-is-it-like-to-be-a-robot/.

D. Graham Burnett, "A Mind in the Water," *Orion Magazine.*

David Clark, "Kant's Aliens," *The New Centennial Review.*

J. M. Coetzee, *The Lives of Animals.*

Simon Critchley, *On Humor.*

Frans de Waal, "Bonobo Sex and Society," *Scientific American.*

Jacques Derrida, *The Animal That Therefore I Am.*

Jacques Derrida, *Of Hospitality.*

Gilles Deleuze, "Difference in Itself," *Difference and Repetition.*

Gilles Deleuze and Félix Guattari, "Introduction: Rhizome" and "1730: Becoming Intense, Becoming-Animal, Becoming-Imperceptable . . . ," in *A Thousand Plateaus.*

Gordon Douglas, dir., *Them!*

Michel Foucault, "Force of Flight," in *Space, Knowledge, and Power: Foucault and Geography.*

Erica Fudge, *Animal.*

Alexander Galloway and Eugene Thacker, *The Exploit: A Theory of Networks.*

Yang Ge, *Strange Beasts of China.*

Lisa-Ann Gershwin, *Stung! On Jellyfish Blooms and the Future of the Ocean.*

William Golding, *The Inheritors.*

Margret Grebowicz, *Whale Song.*

Terike Haapoja and Laura Gustafsson, *History According to Cattle.*

Anders Halverson, *An Entirely Synthetic Fish: How Rainbow Trout Beguiled America and Overran the World.*

Donna Haraway, *Staying with the Trouble: Making Kin in the Chthulhucene.*

Werner Herzog, dir., *Cave of Forgotten Dreams.*

Dougald Hine and Paul Kingsnorth, *Uncivilization: The Dark Mountain Manifesto.*

Jason Hribal, *Fear of the Animal Planet.*

Emir Kusturica, dir., *Underground.*

Francis Lawrance, dir., *I Am Legend.*

Mark Lewis, dir., *Cane Toads.*

John C. Lilly, *Man and Dolphin.*

Alphonso Lingis, *The Imperative.*

H. P. Lovecraft, *At the Mountains of Madness.*

Richard Matheson, *I Am Legend.*

James Mollison, *James and Other Apes.*

Grant Morrison and Frank Quitely, *We3.*

Grant Morrison, Chas Troug, Doug Hazleywood, and Tom Grummett, *Animal Man.*

Fredrich Nietzsche, *The Birth of Tragedy.*

Kira O'Reilly, *inthewrongplaceness.*

Timothy Pachirat, *Every Twelve Seconds: Industrialized Slaughter and the Politics of Sight.*

Elena Passarello, *Animals Strike Curious Poses.*

Dominic Pettman, *Creaturely Love: How Desire Makes More and Less Than Human.*

Mary Roach, *Fuzz: When Nature Breaks the Law.*

Seti-X, *Scrambles of Earth,* https://earthscramble.com/.

Bryndis Snæbjörnsdóttir and Mark Wilson, *You Must Carry Me Now: The Cultural Lives of Endangered Species.*

Isabelle Stengers, "Cosmopolitical Proposal," in *Making Things Public.*

Eugene Thacker, *In the Dust of This Planet.*

Jakob von Uexküll, *A Foray into the Worlds of Animals and Humans.*

Bill Viola, dir., *I Don't Know What It Is That I Am Like.*

George Washington's Mount Vernon, "The Trouble with Teeth," https://www.mountvernon.org/george-washington/health/washingtons-teeth/.

Kari Weil, *Thinking Animals.*

Alan Weisman, *The World Without Us.*

Cary Wolfe, *Before the Law: Humans and Other Animals in a Biopolitical Frame.*

Rupert Wyatt, dir., *Rise of the Planet of the Apes.*

Kate Wyer and Katie Field, *Land Beast.*

Fredrick Young, "Animality," in *Glossalalia.*

Ron Broglio lives in the desert Southwest, where he explores nonhuman phenomenologies. He writes in areas of animal studies, contemporary art, critical theory, and British landscape aesthetics, and he engages in thinking and making with a range of artists who manifest alternative futures.

Media and participatory practice artist **Marina Zurkow** connects people to nature–culture tensions and environmental messes, offering humor and new ways of knowing, connecting, and feeling. Her collaborative initiatives include Climoji, Dear Climate, More&More Unlimited, and Floating Studio for Dark Ecologies. She is working on visualizing future oceans and on connecting eaters to food opportunities in changing climates.

Eugene Thacker is the author of several books, including *In the Dust of This Planet* and *Infinite Resignation*. He teaches at The New School in New York City.